DATE DUE

~~06652042~~	~~12-17-98~~
TLC 41143729	11/24/99
JUN 21 2002	

The RF and Microwave Circuit Design Cookbook

For a complete listing of the *Artech House Microwave Library*,
turn to the back of this book.

The RF and Microwave Circuit Design Cookbook

Stephen A. Maas

Artech House
Boston • London

Library of Congress Cataloging-in-Publication Data
Maas, Stephen A.
The RF and microwave circuit design cookbook / Stephen A. Maas.
p. cm. — (Artech House microwave library)
Includes bibliographical references and index.
ISBN 0-89006-973-5 (alk. paper)
1. Radio circuits. 2. Microwave circuits. I. Title.
II. Series.
TK6560.M22 1998
621.381'32—dc21 98-28219
CIP

British Library Cataloguing in Publication Data
Maas, Stephen A.
The RF and microwave circuit design cookbook. — (Artech House microwave library)
1. Microwave circuits—Design and construction 2. Radio circuits—Design and construction
I. Title
621.3'8132
ISBN 0890069735

Cover design by Elaine Donnelly

685 Canton Street
Norwood, MA 02062

International Standard Book Number: 0-89006-973-5
Library of Congress Catalog Card Number: 98-28219

10 9 8 7 6 5 4 3 2 1

To the memory of my mother,
Dorothy Louise Pierce Maas

Contents

Preface *xvii*

Chapter 1 Microwave Circuits and Circuit Elements 1

1.1 TRANSMISSION LINES DESCRIBED IN THIS CHAPTER 1

1.2 TRANSMISSION LINE THEORY 3

1.2.1 The Ideal LC Transmission Line 3

1.2.2 Propagating Waves on a Transmission Line 9

1.2.3 What Happens at a Discontinuity? 11

1.2.4 Input Impedance 12

1.2.5 Standing Waves, VSWR, and Return Loss 13

1.2.6 Transmission-Line Loss 14

1.2.7 Coupled Lines 15

1.3 PLANAR TRANSMISSION LINES 17

1.3.1 Substrate Materials 17

1.3.2 Characteristics of Planar Transmission Lines 21

1.3.3 Microstrip 23

1.3.4 Coplanar Waveguide 24

1.3.5 Stripline 25

1.3.6 Suspended-Substrate Stripline 27

1.4 CIRCUIT ELEMENTS 28

1.4.1 Resistors 28

1.4.2 Capacitors 29

1.4.3 Inductors 31

1.5 DISTRIBUTED CIRCUIT ELEMENTS 33

1.5.1 Transmission-Line Stubs 33

1.5.2 Radial Stub 35

1.5.3 Series Lines 36
1.5.4 Discontinuities 36
1.6 SCATTERING PARAMETERS 37
1.6.1 Wave Variables 37
1.6.2 Traveling Waves 40
1.6.3 Multiport Scattering Variables 40
1.6.4 Conversions Between Scattering Parameters and Other Parameter Sets 41
1.6.5 Useful Expressions for Two-Ports 43
1.7 MICROWAVE COMPONENTS 43
1.7.1 Hybrid Couplers and Baluns 44
1.7.2 Directional Couplers 46
1.7.3 Circulators and Isolators 48
REFERENCES 49

Chapter 2 Solid-State Devices 51
2.1 SCHOTTKY-BARRIER DIODES 51
2.1.1 Fundamental Properties 53
2.1.2 Electrical Characteristics 54
2.1.3 Practical Schottky Diodes 56
2.1.4 Diode Selection 57
2.1.5 Diodes for Monolithic Circuits 58
2.1.6 Diode Measurements 59
2.2 VARACTOR AND STEP-RECOVERY DIODES 61
2.2.1 Fundamental Properties 61
2.2.2 Electrical Characteristics 61
2.2.3 Equivalent Circuit 62
2.3 BJTS 62
2.3.1 Fundamental Properties 63
2.3.2 Electrical Characteristics 64
2.3.3 BJT Figures of Merit 66
2.3.4 Other Parasitics 67
2.3.5 Large-Signal Equivalent Circuit 67
2.3.6 Small-Signal Equivalent Circuit 67
2.3.7 Gummel-Poon Model 68
2.4 HBTS 73
2.4.1 Fundamental Properties 73
2.4.2 Equivalent Circuit 74
2.5 FET DEVICES 74

2.6 JFETS 75
2.6.1 Structure and Operation 75
2.6.2 Electrical Characteristics 77
2.6.3 Large-Signal Equivalent Circuit 78
2.6.4 Small-Signal Equivalent Circuit 78
2.6.5 Performance Characteristics 79
2.7 GaAs MESFETS 81
2.7.1 Structure and Operation 81
2.7.2 Device Size and Geometry 83
2.7.3 Electrical Characteristics 84
2.7.4 Large-Signal Equivalent Circuit 84
2.7.5 Small-Signal Equivalent Circuit 85
2.7.6 Performance Characteristics 85
2.8 HEMTS 87
2.8.1 Structure and Operation 87
2.8.2 Device Size and Geometry 88
2.8.3 Electrical Characteristics 89
2.8.4 Large-Signal Equivalent Circuit 89
2.8.5 Small-Signal Equivalent Circuit 89
2.8.6 Performance Characteristics 89
2.9 MOSFETs 90
2.9.1 Structure and Operation 90
2.9.2 Device Size and Geometry 90
2.9.3 Electrical Characteristics 91
2.9.4 Large-Signal Equivalent Circuit 92
2.9.5 Small-Signal Equivalent Circuit 92
2.9.6 Performance Characteristics 92
REFERENCES 93

Chapter 3 Diode Mixers 95
3.1 DIODE MIXER THEORY AND OPERATION 95
3.1.1 Fundamentals 95
3.1.2 Other Important Performance Characteristics 100
3.1.3 Balanced Mixers 101
3.1.4 Baluns and Hybrids 101
3.2 SINGLY BALANCED, 180-DEGREE “RAT-RACE” MIXER 103
3.2.1 Characteristics 103
3.2.2 Description 103

3.2.3 Design 104
3.2.4 Variations 105
3.2.5 Cautions 107
3.3 SINGLY BALANCED, 90-DEGREE MIXER 107
3.3.1 Characteristics 107
3.3.2 Description 108
3.3.3 Design 108
3.3.4 Variations 109
3.3.5 Cautions 109
3.4 DOUBLY BALANCED RING MIXER USING COUPLED-LINE BALUNS 109
3.4.1 Characteristics 109
3.4.2 Description 109
3.4.3 Design 110
3.4.4 Variations 113
3.4.5 Cautions 114
3.5 DOUBLY BALANCED "HORSESHOE" BALUN MIXER 115
3.5.1 Characteristics 115
3.5.2 Description 115
3.5.3 Design 115
3.5.4 Variations 118
3.5.5 Cautions 118
3.6 DOUBLY BALANCED STAR MIXER 119
3.6.1 Characteristics 119
3.6.2 Description 119
3.6.3 Design 121
3.6.4 Variations 122
3.6.5 Cautions 122
3.7 MONOLITHIC CIRCUITS 123
REFERENCES 124

Chapter 4 Diode Frequency Multipliers 125

4.1 FREQUENCY-MULTIPLIER THEORY 125
4.1.1 Resistive Frequency Multipliers 127
4.1.2 Varactor Multipliers 129
4.1.3 Step-Recovery-Diode Multipliers 134
4.2 SINGLE-DIODE RESISTIVE FREQUENCY DOUBLER 138
4.2.1 Characteristics 138

4.2.2 Description 138
4.2.3 Design 139
4.2.4 Variations 140
4.2.5 Cautions 140
4.3 SINGLY BALANCED FREQUENCY DOUBLER: INPUT BALUN 141
4.3.1 Characteristics 141
4.3.2 Description 141
4.3.3 Design: Rat-Race Multiplier 142
4.3.4 Design: Coplanar Multiplier 144
4.3.5 Variations 145
4.3.6 Cautions 145
4.4 SINGLY BALANCED DOUBLER: OUTPUT BALUN 145
4.4.1 Characteristics 145
4.4.2 Description 146
4.4.3 Design 147
4.4.4 Variations 148
4.4.5 Cautions 148
4.5 DOUBLY BALANCED RESISTIVE FREQUENCY DOUBLER 148
4.5.1 Characteristics 148
4.5.2 Description 149
4.5.3 Design 150
4.5.4 Variations 150
4.5.5 Cautions 151
4.6 VARACTOR FREQUENCY MULTIPLIERS 151
4.6.1 Characteristics 151
4.6.2 Description 152
4.6.3 Design 152
4.6.4 Variations 154
4.6.5 Cautions 154
4.7 STEP-RECOVERY-DIODE FREQUENCY MULTIPLIER 154
4.7.1 Characteristics 154
4.7.2 Description 155
4.7.3 Design 155
4.7.4 Variations 157
4.7.5 Cautions 157
REFERENCES 157

Chapter 5 Other Diode Applications 159

5.1 DIODE DETECTORS 159
5.1.1 Fundamental Properties 160
5.1.2 Detector Diodes 161
5.1.3 Square-Law Detection 162
5.1.4 Envelope Detection 163

5.2 SQUARE-LAW DETECTORS 164
5.2.1 Characteristics 164
5.2.2 Description 164
5.2.3 Design 164
5.2.4 Cautions 166
5.2.5 Variations 166

5.3 ENVELOPE DETECTORS 166
5.3.1 Characteristics 166
5.3.2 Design 167
5.3.3 Variations 167
5.3.4 Cautions 168

5.4 DOUBLE-SIDEBAND (DSB) MODULATORS 168
5.4.1 Characteristics 168
5.4.2 Description 169
5.4.3 Design 169
5.4.4 Variations 170
5.4.5 Cautions 170

5.5 SINGLE-SIDEBAND (SSB) MODULATORS 171
5.5.1 Characteristics 171
5.5.2 Description 171
5.5.3 Design 172
5.5.4 Variations 172
5.5.5 Cautions 172

5.6 I-Q MODULATORS 173
5.6.1 Characteristics 173
5.6.2 Description 173
5.6.3 Design 174
5.6.4 Variations 174

REFERENCES 174

Chapter 6 Active Mixers 175

6.1 ACTIVE MIXER THEORY 175

6.1.1 Transconductance Mixers 177
6.1.2 Conversion Efficiency 180
6.1.3 Single-Device Equivalent Circuit 181
6.1.4 Other Configurations 181
6.2 SINGLE-FET MICROWAVE MIXER 182
6.2.1 Characteristics 182
6.2.2 Description 183
6.2.3 Design 184
6.2.4 Variations 188
6.2.5 Cautions 189
6.3 SINGLE-DEVICE, DUAL-GATE MIXER 190
6.3.1 Characteristics 190
6.3.2 Description 190
6.3.3 Design 192
6.3.4 Design Example 195
6.3.5 Variations 199
6.3.6 Cautions 199
6.4 SINGLY BALANCED FET MIXERS 199
6.4.1 Characteristics 199
6.4.2 Description 200
6.4.3 Design 201
6.4.4 Variations 202
6.4.5 Cautions 202
6.5 SINGLY BALANCED DIFFERENTIAL MIXER 202
6.5.1 Characteristics 202
6.5.2 Description 202
6.5.3 Design 203
6.5.4 Variations 205
6.5.5 Cautions 205
6.6 DOUBLY BALANCED MOSFET MIXER 205
6.6.1 Characteristics 206
6.6.2 Description 206
6.6.3 Design 206
6.6.4 Variations 207
6.6.5 Cautions 207
6.7 GILBERT-CELL BIPOLAR MIXER 209
6.7.1 Characteristics 209
6.7.2 Description 210
6.7.3 Design 210

6.7.4 Cautions 211
REFERENCES 211

Chapter 7 FET Resistive Mixers 213

7.1 FUNDAMENTALS OF FET RESISTIVE MIXERS 213
7.1.1 Linear Mixing 214
7.1.2 Our First Linear Mixer: The Hamster-Pumped Mixer 215
7.1.3 An Improved Linear Mixer: The FET Resistive Mixer 216

7.2 SINGLE-DEVICE MIXER: RF APPLICATIONS 221
7.2.1 Characteristics 221
7.2.2 Description 222
7.2.3 Design 223
7.2.4 Variations 225
7.2.5 Cautions 227

7.3 180-DEGREE SINGLY BALANCED FET RESISTIVE MIXER 227
7.3.1 Characteristics 227
7.3.2 Description 227
7.3.3 Design 229
7.3.4 Variations 229
7.3.5 Cautions 232

7.4 DOUBLY BALANCED RING MIXER 232
7.4.1 Characteristics 232
7.4.2 Description 232
7.4.3 Design 234
7.4.4 Variations 236
7.4.5 Cautions 236

7.5 SUBHARMONICALLY PUMPED FET RESISTIVE MIXER 236
7.5.1 Characteristics 236
7.5.2 Description 238
7.5.3 Design 239
7.5.4 Variations 239
7.5.5 Cautions 239
REFERENCES 240

Chapter 8 Active Frequency Multipliers 241

8.1 ACTIVE FREQUENCY-MULTIPLIER THEORY 241
8.1.1 Why Use Active Multipliers? 241
8.1.2 Active Multiplier Operation 242

8.2 SINGLE-DEVICE MICROWAVE FREQUENCY MULTIPLIER 249
8.2.1 Characteristics 249
8.2.2 Description 250
8.2.3 Design 250
8.2.4 Variations 253
8.2.5 Cautions 254
8.3 BALANCED FREQUENCY DOUBLER 254
8.3.1 Characteristics 254
8.3.2 Description 255
8.3.3 Design 256
8.3.4 Variations 257
8.3.5 Cautions 257
REFERENCES 259

About the Author 261

Index 263

Preface

If you are an academic, stop reading this book and put it down right now. Go back to your office, have a nice cup of tea, and review one of those papers you set aside six weeks ago. If you do this, you'll feel better. If you keep reading, you will just get upset. This book is not for you, anyway.

Quite a long time ago, when I was a tender, young undergraduate at the University of Pennsylvania (OK, although I never was tender and have a hard time believing that I was ever young; there is documentary evidence that I was an undergrad, however, so I believe this story), I took a course in engineering math from an electromagnetics professor. At one point, he turned to the blackboard to begin his exposition, then half turned back to the class and said, "We must study this material. It is important! After all, *we* don't want to become hardware engineers!"

This book is my separate peace as a full member of the crowd for whom that professor showed so much disdain. Most of us are practical, hardware engineers, and we know a lot more than the academics think we do. Furthermore, that knowledge is hard to obtain, probably harder than theoretical knowledge. After all, there are college classrooms full of students learning electromagnetic theory, but none where those students learn how to avoid ground problems. Too trivial, I suppose.

It's not too much of an exaggeration to say that most technical books are written for people who don't need them. They are written by academics for people like themselves. Who is looking after the rest of us? Certainly, precious few books address the practical aspects of engineering and technology, yet both new and experienced engineers continually ask for them. I have always admired those few that do: Matthaei, Young, and Jones' book on filters, Wadell's *Transmission Line Design Handbook*, and Press, Flannery, Teukolsy, and Vetterling's *Numerical Recipes*. I dare anyone to accuse these authors of triviality or superficiality. While most texts dump a huge amount of theory on the reader, and leave him alone to figure out what to do with it, these texts emphasize the use of the material, while giving the reader just enough theory so he knows what he's doing. With this book,

I'm staking out a claim in the same territory.

This book is designed primarily for engineers in their first few years of practice, as they struggle to develop some new skills while under pressure to build hardware and get it out the door. Perhaps this will bootstrap the process of gaining experience and will ease their struggle with their first few designs. More experienced engineers may know much of the material presented here. Still, there are things in this book that took me years to learn, so I hope that almost everyone will find something useful.

The first two chapters cover basic theory of solid-state devices and circuit structures. They are not intended to be exhaustive; if they were, each would be a book in itself. Instead, they are designed to present the most important aspects of the material necessary for designing microwave circuits. The remaining chapters are organized similarly. The first section covers basic theory, and the following sections describe the design of a single type of circuit. These are organized into five subsections: "Characteristics," the properties of the circuit; "Description," the circuit itself; "Design," the design procedure, as specific and "cookbook-like" as possible; "Variations," other useful modifications of the circuit; and "Cautions," pitfalls in the design process or in the circuit's implementation.

Each chapter begins with the description of a single-device version of the circuit. Even if a single-device circuit is not what you want, it's a good idea to read this section before the others. Single-device circuits are prototypes for balanced circuits, and the descriptions of multiple-device circuits later in the chapter make frequent references to the single-device circuit.

This cookbook approach has the obvious advantage of simplifying the design process. By being specific instead of general, however, it has a potentially serious disadvantage: it fails to address the variety of design approaches that a broad understanding of the technology allows. Indeed, developing and using this kind of broad technical knowledge is really what engineering is all about, and we purposely sidestep it here. Is that a good idea? I'll be so bold as to defend it. We all have to start somewhere, and by explaining the process of designing certain specific circuits, perhaps this approach may show the beginner the underlying logic. He then can apply that logic to a much broader range of designs. In the long run, this may help him develop the broad-range design skills faster and more completely than a focus solely on theory.

Steve Maas
Nonlinear Technologies, Inc.
Long Beach, California

Chapter 1
Microwave Circuits and Circuit Elements

The main thing that distinguishes microwave and radio-frequency (RF) circuits from lower-frequency circuits is the need to include distributed effects. In low-frequency circuits, inductors are inductors, capacitors are capacitors, resistors are resistors, and wires, no matter how long, are simply nodes. Not so in the RF and microwave world.

In high-frequency circuits, capacitors and inductors often are realized by transmission-line segments. Transmission lines often must be used for circuit interconnections as well. Even when lumped circuit elements are employed, transmission-line segments may be needed to model them accurately. Clearly, before we can do much of anything useful, we need to deal with the subject of transmission lines.

1.1 TRANSMISSION LINES DESCRIBED IN THIS CHAPTER

This chapter is concerned primarily with planar transmission lines (flat conductors on a dielectric substrate), since these are the most practical for use in the components considered in later chapters. Table 1.1 shows the types of transmission lines described in this chapter. We begin with a review of transmission-line theory in Section 1.2. A description of the properties of planar transmission lines and substrates begins in Section 1.3.

We concentrate on these few structures because they are fundamental. Many variations are possible. For example, microstrip can be placed in an enclosure ("microstrip in a box") or can have a cover. Coplanar waveguide (CPW) can have an additional ground plane under the substrate or a ground plane on only one side of the conductor. These variants have many of the same basic characteristics as the fundamental structure from which they are derived.

For further information on such structures, see Wadell [1].

Table 1.1 Planar Transmission Lines Described in This Chapter

Transmission Line	Structure	Properties
Microstrip	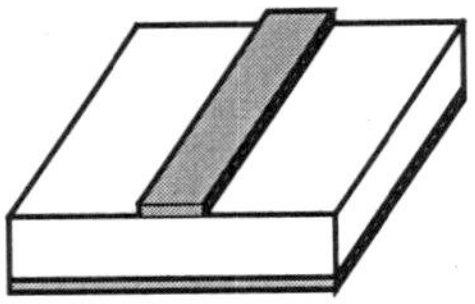	The most common type of transmission line, suitable for both hybrids and monolithic circuits. Moderately dispersive at high frequencies. See Section 1.3.3.
Coplanar waveguide (CPW)		Somewhat lossier and more dispersive than microstrip, but minimizes the parasitic inductance of ground connections. Good transition to coaxial lines. Spurious slotline and microstrip modes are possible. See Section 1.3.4.
Stripline	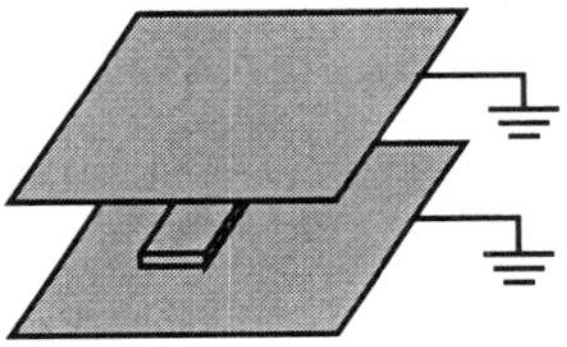	Does not allow convenient mounting of discrete circuit elements; best for passive components. Difficult to cascade with microstrip or other planar transmission lines. Low loss, TEM, good transition to coax. See Section 1.3.5.
Suspended-substrate stripline (SSSL)	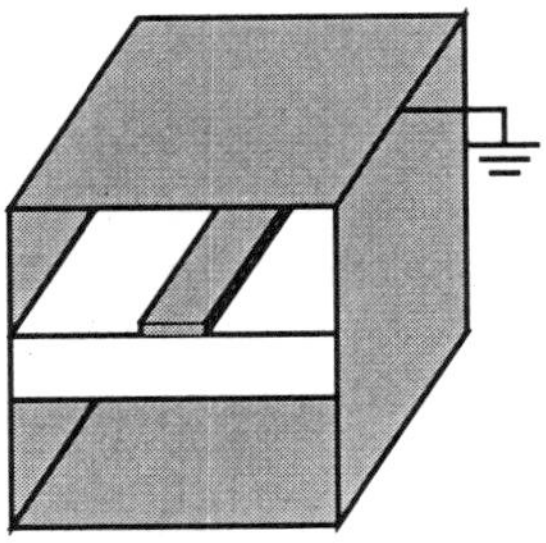	Similar to stripline, but easier to fabricate in many types of circuits. Low loss, low effective dielectric constant, good transition to coax. Waveguide-like modes can be a problem. See Section 1.3.6.

1.2 TRANSMISSION LINE THEORY

Left to its own devices, an electromagnetic wave propagates in a straight line. Often we don't like this. Getting a wave to go where we want it to go, as long as it's some direction other than a straight line, is the job of a transmission line.

There are a number of ways to approach the subject of transmission lines. One is to solve Maxwell's equations consistent with boundary conditions imposed by the structure itself. This is necessary for certain types of structures (cylindrical or rectangular waveguides, for example), but it isn't very general. A more general treatment views the transmission-line's conductors as uniform structures having series inductance, shunt capacitance, and perhaps resistance to account for the transmission line's loss. This approach is applicable to a wide variety of transmission lines. Determining the inductance and capacitance of a particular type of line is, of course, a problem in electromagnetics.

1.2.1 The Ideal LC Transmission Line

Consider the infinite cascade of LC sections shown in Figure 1.1. This is a low-pass structure having the cutoff frequency

$$\omega_c = \frac{2}{\sqrt{LC}} \tag{1.1}$$

where L and C are the inductance and capacitance per section. The phase shift of a sinusoidal signal propagating on the cascade is

$$\phi = 2\,\mathrm{asin}\left(\frac{\omega}{2}\sqrt{LC}\right) \tag{1.2}$$

and the terminating impedance giving this result is

$$Z_t = \sqrt{\frac{L}{C}} \tag{1.3}$$

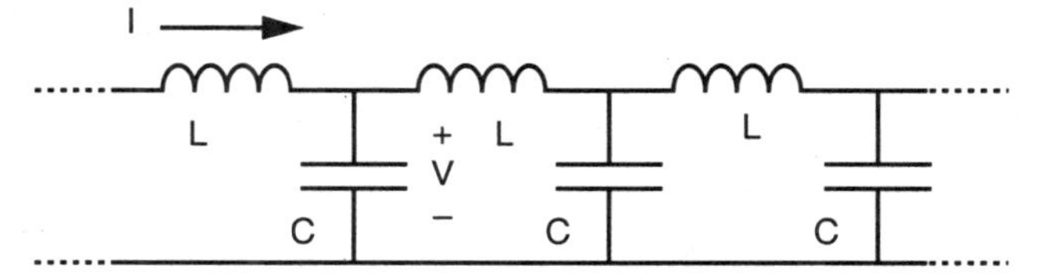

Figure 1.1 A transmission line is modeled as a cascade of series inductance and shunt capacitance sections.

Now imagine that the values of L and C are reduced, keeping the ratio L/C constant, and the number of sections is increased so that the capacitance and inductance per meter are kept constant as $L, C \rightarrow 0$. Clearly, the terminating impedance remains the same, but $\omega_c \rightarrow \infty$. This means that a signal can propagate unattenuated from section to section, at any frequency. As $L, C \rightarrow 0$, the phase shift per section becomes

$$\phi = \omega\sqrt{LC} \tag{1.4}$$

which obviously approaches zero. However, since the total capacitance and inductance per unit of length are constant, we can say

$$\phi_l = \omega\sqrt{L_l C_l} \tag{1.5}$$

where the subscript l indicates the quantity per meter. The phase shift per meter is simply the radian frequency divided by the phase velocity, so

$$\begin{aligned} \omega\sqrt{L_l C_l} &= \frac{\omega}{v_p} \\ v_p &= \frac{1}{\sqrt{L_l C_l}} \end{aligned} \tag{1.6}$$

indicating that a signal can propagate happily along this line unattenuated and at a velocity that is independent of its frequency. Finally, the appropriate terminating impedance, which we now shall call the *characteristic impedance* of the line[1], is

$$Z_0 = \sqrt{\frac{L_l}{C_l}} \tag{1.7}$$

Although this result is based on sinusoidal signals, it has an important implication for general, nonsinusoidal signals. From Fourier analysis we know that any such signal can be viewed as the sum of a number (sometimes an infinite number) of sinusoids. If all the frequency components of such a signal are unattenuated and propagate at the same velocity, the signal itself also must be undistorted as it propagates along the line. Thus, the above results are valid for any signal, and the line is distortionless.

We now need to face the problem of determining L_l and C_l for a practical transmission line. In many cases this is a difficult problem, and a delightful source of productive labor for academic electromagneticists. In others, however, it is relatively simple. The task is best illustrated by an example.

---

1. Relax. We define this term more precisely on page 10.

Example 1: Coaxial Line

One of the most practical transmission lines consists of a metallic tube, called the *outer conductor*, and a concentric *center conductor*. The center conductor is usually supported by a solid dielectric. The electric field is radial and the magnetic field is circumferential. Figure 1.2 shows the geometry of such a line.

As with all transmission lines, we can view its operation in two ways. One is to consider the currents in the conductors and the voltage between them. This gives us a view that is comfortably consistent with the discussion so far and is easily identifiable with Figure 1.1. On the other hand, we can view the transmission line as a type of wave-guiding structure and focus on the wave nestled comfortably between the two concentric conductors. Either view works, as long as it is properly conceived. Both views tell us something complementary about the way the transmission line works, so it is especially helpful to be able to switch easily between them. We do a lot of this in the following discussion.

We first consider the capacitance between the inner and outer conductors. The center conductor has a uniform charge per unit length, Q_l, which is balanced by an equal charge of opposite sign on the inside of the outer conductor. From Gauss's law, which states that the electric flux must equal the enclosed charge, the electric field, E_r, in the region between the conductors must be

$$E_r = \frac{Q_l}{2\pi\varepsilon} \tag{1.8}$$

where ε is the electric permittivity of the dielectric material in the region between the conductors. ε is sometimes written $k\varepsilon_0$, where ε_0 is the permittivity of free space, $8.854 \cdot 10^{-12}$ F/m, and k is the dielectric constant. The voltage V between the inner and outer conductors is

$$V = \int_a^b E_r dr = \int_a^b \frac{Q_l}{2\pi\varepsilon} dr = \frac{Q_l}{2\pi\varepsilon} \ln\left(\frac{b}{a}\right) \tag{1.9}$$

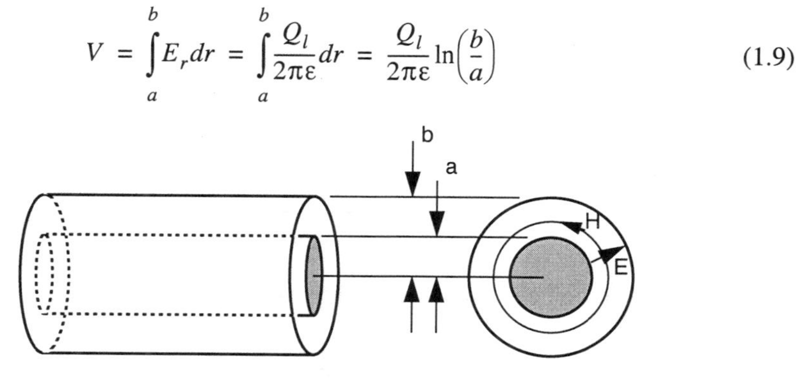

Figure 1.2 A coaxial transmission line consists of two concentric conductors. A wave is guided along the region between the conductors. The electric field, E, is entirely radial and the magnetic field, H, has only circumferential components.

Finally, the capacitance per unit of length is

$$C_l = \frac{Q_l}{V} = \frac{2\pi\varepsilon}{\ln\left(\frac{b}{a}\right)} \tag{1.10}$$

Now for the inductance. Here we use a sneaky trick to avoid a lot of work. From (1.6) we find that

$$L_l = \frac{1}{v_p^2 C_l} \tag{1.11}$$

The region between the conductors contains a plane wave; its phase velocity must be

$$v_p = \frac{c}{\sqrt{k}} \tag{1.12}$$

where c is the velocity of light and k is the dielectric constant of the material in the space between the conductors. Substituting (1.12) into (1.11) gives us the inductance.

This trick is very useful. Even in complex structures, we can find the inductance by setting all dielectric constants equal to 1.0, calculating the capacitance, and applying (1.11) with $v_p = c$. This way, finding L_l and C_l requires analyzing the structure twice, once with $\varepsilon = \varepsilon_0$ and once with $\varepsilon = k\varepsilon_0$. Since only a constant is different, these two analyses are essentially the same, and the approach is much easier than creating separate analyses for the inductance and the capacitance.

Finally, substituting (1.11) into (1.7) gives

$$Z_0 = \frac{1}{v_p C_l} \tag{1.13}$$

or

$$Z_0 = \frac{\ln\left(\frac{b}{a}\right)}{2\pi c\varepsilon_0\sqrt{k}} \tag{1.14}$$

We now obtain some help from two well-known relations,

$$c = \frac{1}{\sqrt{\mu_0\varepsilon_0}} \tag{1.15}$$

and

$$\eta_0 = \sqrt{\frac{\mu_0}{\varepsilon_0}} \tag{1.16}$$

where μ_0 is the permeability of free space, $4\pi \cdot 10^{-7}$ H/m. With these two relations, (1.14) can be manipulated into the convenient form

$$Z_0 = \frac{\eta_0}{2\pi\sqrt{k}} \log\left(\frac{b}{a}\right) = \frac{138}{\sqrt{k}} \log\left(\frac{b}{a}\right) \tag{1.17}$$

where *log* is the base 10 logarithm. If $k = 1.0$, a 50Ω transmission line requires $b/a = 2.30$. This is a nice number to remember.

It would be pleasant[2] if all transmission-line analyses were so simple. Unfortunately, for most types of transmission lines it is not possible to find simple, exact, closed-form expressions like (1.17). Furthermore, this approach obscures a dirty little secret about all transmission lines: we have quietly assumed that the wave in the region between the inner and outer conductors is a transverse electromagnetic (TEM) wave, in effect a plane wave whose electric and magnetic fields are perpendicular to the axis of the coaxial line. Although this is valid for most practical coaxial lines, it isn't always the case. As soon as the width of the region between the conductors approaches one half wavelength, other field structures (called *modes*) can propagate, and the characteristic impedances and phase velocities of the various modes are different. The effect of mode generation is sometimes minor but often serious. Components that depend on a purely TEM mode for proper operation (filters, for example) may operate poorly when unwanted modes are present.

An approximate expression for the cutoff frequency of the first non-TEM mode in coaxial line is

$$f_c \approx \frac{191}{\sqrt{k}(a+b)} \tag{1.18}$$

where a and b are in millimeters and f_c is in GHz. This expression implies that, as with other types of transmission lines, modes are a problem only when we try to operate relatively large transmission lines at very high frequencies.

Example 2: Dispersion

Another phenomenon to worry about is dispersion. To illustrate this, suppose that the coaxial line in Figure 1.2 is loaded with two different dielectrics, as shown in Figure 1.3. Specifically, we assume that the dielectric has the value $\varepsilon_1 = k_1\varepsilon_0$ from radius a

2. For us, that is; for electromagnetics Ph.D. students in need of a dissertation topic, it would be a disaster.

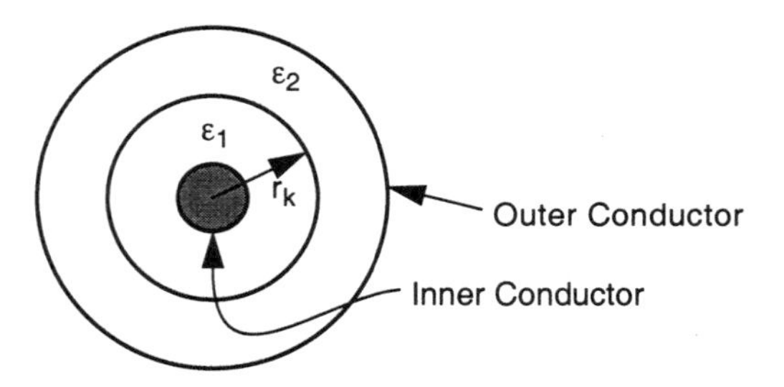

Figure 1.3 End view of a coaxial transmission line containing two different dielectrics.

to r_k and $\varepsilon_2 = k_2\varepsilon_0$ from radius r_k to b. The expression for voltage becomes, instead of (1.9),

$$V = \int_a^{r_k} \frac{Q_l}{2\pi\varepsilon_1} dr + \int_{r_k}^{b} \frac{Q_l}{2\pi\varepsilon_2} dr \tag{1.19}$$

The resulting capacitance is

$$C_l = \frac{2\pi\varepsilon_0}{\frac{1}{k_1}\ln\left(\frac{r_k}{a}\right) + \frac{1}{k_2}\ln\left(\frac{b}{r_k}\right)} \tag{1.20}$$

Once we determine the phase velocity, the inductance and characteristic impedance follow as before.

Now, what about the phase velocity? Since the dielectric is not homogeneous, a simple equation like (1.12) clearly is not valid. However, the inductance is independent of the dielectric, so

$$L_l = \frac{1}{v_p^2 C_l} = \frac{1}{c^2 C_0} \tag{1.21}$$

where C_0 is the capacitance of the structure per unit length when $k_1 = k_2 = 1.0$. From (1.21) we find that

$$v_p = c\sqrt{\frac{C_0}{C_l}} \tag{1.22}$$

This result is simple and elegant. Too bad it's wrong! The problem lies in the assumption that the region between the conductors supports a TEM wave. In reality, a TEM wave is impossible in such a structure, because it would have to propagate at different velocities in the two dielectrics, a clear impossibility. To satisfy Maxwell's equations in such a structure, the field must have longitudinal components. Those components depend on frequency, so the frequency-independent velocity implied by (1.22) is not possible.

Because it is based on electrostatic calculations, (1.22) is strictly correct only as frequency approaches zero; for this reason the derivation is sometimes called a *quasistatic analysis*. In essence, we have assumed that the line's characteristics are the same at high frequencies as at dc. At high frequencies the phase velocity deviates from the value given by (1.22). One effect of this frequency-dependent phase velocity is to distort a complex waveform as it propagates down the transmission line. If the waveform is a pulse, for example, dispersion "smears" it out in time, slowing the rise and fall times. This phenomenon is appropriately called *dispersion*.

Dispersion depends on the frequency, the diameters of the inner and outer conductors, and the ratio of the dielectric constants. If the dielectric discontinuity is small and the diameters a and b are small relative to a wavelength, dispersion often can be neglected and (1.22) is accurate. The problem faced by a designer is to know when the TEM assumption is valid and when it isn't. This is less of a problem with coaxial lines, which rarely use nonhomogeneous dielectrics, than in microstrip or other planar transmission lines, which are nonhomogeneous by their very nature. In the latter types of lines, determining the need for dispersion corrections and making appropriate corrections when they are warranted are important parts of the design process.

1.2.2 Propagating Waves on a Transmission Line

It's time to be a little more specific about waves on transmission lines. The current I and voltage V on the line (Figure 1.1) satisfy the equations

$$\begin{aligned} \frac{\partial V}{\partial z} &= -L_l \frac{\partial I}{\partial t} \\ \frac{\partial I}{\partial z} &= -C_l \frac{\partial V}{\partial t} \end{aligned} \tag{1.23}$$

where z is the longitudinal position on the line. Differentiating both equations with respect to t and z and equating them give the telegrapher's equations,

$$\begin{aligned} \frac{\partial^2 V}{\partial z^2} &= L_l C_l \frac{\partial^2 I}{\partial t^2} \\ \frac{\partial^2 I}{\partial z^2} &= L_l C_l \frac{\partial^2 V}{\partial t^2} \end{aligned} \tag{1.24}$$

The solution is

$$V(t, z) = f_1\left(t - \frac{z}{v_p}\right) + f_2\left(t + \frac{z}{v_p}\right)$$
$$I(t, z) = g_1\left(t - \frac{z}{v_p}\right) + g_2\left(t + \frac{z}{v_p}\right) \tag{1.25}$$

where f_1, f_2, g_1, and g_2 are arbitrary, continuous, real functions and $v_p = 1/\sqrt{L_l C_l}$. At first glance it seems surprising that an arbitrary function should satisfy these equations, but in reality it simply means that any waveform can propagate on the line. The function's argument shows that the function is displaced in time by z/v_p as it propagates down the line, f_1 in the $+z$ direction and f_2 in the $-z$ direction. In other words, any waveform can propagate undistorted in either direction on an ideal transmission line at a velocity $v_p = 1/\sqrt{L_l C_l}$. What a revelation!

We are especially interested in the case where f_1 and f_2 are sinusoidal functions. Expressing these sinusoids as phasors, we have, for the voltage wave,

$$V = V_f \exp(-\gamma z) + V_r \exp(\gamma z) \tag{1.26}$$

where V_f and V_r are the forward- and reverse-propagating waves, respectively; the $j\omega$ dependence is not explicit; and γ is the propagation constant.

$$\gamma = j\beta$$
$$\beta = \frac{\omega}{v_p} \tag{1.27}$$

β is the phase shift per unit of distance along the line. The expression for the current is analogous.

We are now in a position to describe the characteristic impedance more precisely. Substituting expressions for the forward-propagating current and voltage waves into the first equation of (1.23) gives

$$-\gamma V_f = -L_l j\omega I_f \tag{1.28}$$

Substituting (1.27) gives

$$\frac{V_f}{I_f} = \sqrt{\frac{L_l}{C_l}} \tag{1.29}$$

which looks a lot like (1.7). In other words, the characteristic impedance, Z_0, is the ratio of voltage to current in the propagating wave on the line. Of course, the same relation applies to the reverse-propagating waves.

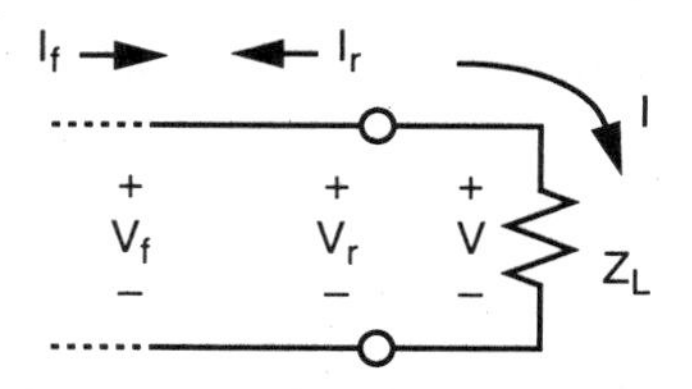

Figure 1.4 At the load interface of the transmission line, the forward- and reverse-propagating waves' voltages and currents must satisfy Kirchoff's laws.

1.2.3 What Happens at a Discontinuity?

Normally we use a transmission line to send a signal from one place to another, and we are distressed at the possibility that part of our signal might be reflected by the load and returned to its place of origin. For better or worse, (1.25) shows this to be a real possibility. What conditions might cause it actually to happen?

Figure 1.4 shows the interface between a transmission line and a load. If we have only a forward-propagating wave, there is a clear problem: at the interface Kirchoff's laws are violated, since $I_f = I$ and $V_f = V$, but $V_f / I_f = Z_0$ and $V / I = Z_L$. To satisfy Kirchoff's laws at the interface there must be a reverse-propagating wave. Then

$$\frac{V_f + V_r}{I_f - I_r} = \frac{V}{I} = Z_L \tag{1.30}$$

and

$$\begin{aligned} \frac{V_f}{I_f} &= Z_0 \\ \frac{V_r}{I_r} &= Z_0 \end{aligned} \tag{1.31}$$

as before. This can be solved, although the algebra is a little sticky. The result is

$$\Gamma = \frac{V_r}{V_f} = \frac{I_r}{I_f} = \frac{Z_L - Z_0}{Z_L + Z_0} \tag{1.32}$$

where Γ is the reflection coefficient. If you are wondering why we chose different reference directions for I_f and I_r, there is a good reason: if we had made the reference directions the same, the current and voltage reflection coefficients would be the negative of each other. Most people feel that this arrangement is more convenient.

A few reflection coefficients are worth remembering. When the load is matched,

$Z_L = Z_0$ and the reflection coefficient is zero. A short-circuit load has $\Gamma = -1.0$ and an open circuit has $\Gamma = 1.0$.

1.2.4 Input Impedance

Inevitably we need to know the input impedance of the line,

$$Z_{in} = \frac{V_{in}}{I_{in}} \tag{1.33}$$

where V_{in} and I_{in} are the voltage and the current at the terminals of the line where the excitation is applied. Once we know the reflection coefficient, the input impedance of the line is easy to determine. As we move l meters away from the load and toward the excitation source, the phase of the forward wave is advanced by γl and the reflected wave is delayed by the same quantity. The resulting expression for the input reflection coefficient is

$$\Gamma_{in} = \Gamma_L \exp(-2j\gamma l) \tag{1.34}$$

where Γ_{in} is the input reflection coefficient and Γ_L is the load reflection coefficient. On an ideal line the magnitude of the reflection coefficient does not change with position; only the phase changes. To find the input impedance, we first recognize that (1.32), although formulated for the load, is valid for any point in the line. Then, by inverting we obtain

$$Z_{in} = Z_0 \frac{1 + \Gamma_{in}}{1 - \Gamma_{in}} \tag{1.35}$$

The explicit formula is

$$Z_{in} = Z_0 \frac{Z_L \cos(\beta l) + jZ_0 \sin(\beta l)}{Z_0 \cos(\beta l) - jZ_L \sin(\beta l)} \tag{1.36}$$

A few interesting facts can be extracted from (1.36):

- A quarter-wavelength, shorted transmission line looks like an open circuit at its input;
- A quarter-wavelength, open-circuited transmission line looks like a short circuit at its input;
- A shorted line less than one-quarter wavelength long is inductive;
- An open-circuited line less than one-quarter wavelength long is capacitive;
- A line that is an integer number of half-wavelengths long has $Z_{in} = Z_L$.

These little gems of information can be used to make very nice inductors, capacitors, and resonators out of short- or open-circuited transmission lines. (We consider this possibility in Section 1.5.)

Knowing the input impedance, we can make sure that all the available power of our source is delivered to the load. We simply conjugate-match the input impedance of the line. Because the line is lossless, all available power must be delivered to the load. In practice, conjugate-matching the input creates a situation where the reflected wave is re-reflected from the input with just the right magnitude and phase for it eventually to deliver all power to the load. We say "eventually" because it may take a few reflective round trips along the line before all the power ends up in the load. For narrowband signals on ideal, lossless lines, this is not much of a problem. However, in broadband systems with real, lossy, dispersive lines, the signal loses power and becomes more distorted with each trip along the line. We therefore try to avoid the use of dispersive transmission lines.

1.2.5 Standing Waves, VSWR, and Return Loss

Let's think a little more deeply about the forward-propagating wave, the load, and the reflected wave. The phase difference between the voltage of the forward wave and the reflected wave, at any point in the line, is the sum of three components:

1. The phase shift, βl, that the forward wave undergoes between the point on the line and the load;
2. The angle of the reflection coefficient;
3. The phase shift, again βl, that the reflected wave undergoes between the load and the point on the line.

These phase shifts do not vary with time, so the phase difference between the voltage of the forward-propagating wave and the reflected wave is constant. At some points on the line, the voltages of the forward and reflected waves are in phase, resulting in a high voltage; at these points, the voltage is

$$V_{max} = |V_f| + |V_r| \tag{1.37}$$

At other points, the waves have a 180-degree phase difference, and the voltage reaches a minimum,

$$V_{min} = |V_f| - |V_r| \tag{1.38}$$

This pattern of maxima and minima repeats every half wavelength along the line. Although it is not really a wave, in the classical sense, this pattern of high and low voltages is called a *standing wave*. The line also has standing waves of current. The current minima coincide with the voltage maxima, and the current maxima coincide with the voltage minima.

The voltage standing wave ratio (VSWR) is the ratio of the magnitudes of maximum and minimum voltage along the line.

$$VSWR = \frac{V_{max}}{V_{min}} = \frac{1 + \dfrac{|V_r|}{|V_f|}}{1 - \dfrac{|V_r|}{|V_f|}} = \frac{1 + |\Gamma|}{1 - |\Gamma|} \tag{1.39}$$

The current standing wave ratio has the same value. (Prove it to yourself!) This brings up the question of the need for a *V* in *VSWR*. It is clearly unnecessary, and sometimes the simplified expression *SWR* is used. The persistence of the *V*, in the absence of any real need for it, must remain one of the great mysteries of electrical science. Even more perplexing is the fact that most people using the term *VSWR* don't care at all about the voltages on the line; they use VSWR simply as an alternate way of stating the magnitude of a reflection coefficient.

Return loss is an even sillier concept. Return loss, *RL,* is

$$RL = 20 \log\left(\left|\frac{V_r}{V_f}\right|\right) = 20 \log(|\Gamma|) \tag{1.40}$$

that is, the power "lost" in the load between the incident and the reflected waves. The main reason for the use of this quantity is that engineers like to express everything in decibels. Most have trouble dealing with a scalar quantity.

1.2.6 Transmission-Line Loss

Every thirteen-year-old amateur radio operator knows that some of the power his transmitter pumps into his transmission line does not reach his antenna. It is dissipated in resistive losses in the line. These resistances are generally greater at high frequencies than at dc, because as frequency increases, the current in the transmission line is concentrated in a progressively thinner region near the surface of the conductors. This phenomenon is called *skin effect*. The dielectric that insulates the transmission line's conductors also may introduce loss. This loss may arise in the dielectric's finite bulk resistance, but more often it results from molecular resonances that absorb energy and mimic conduction. In most practical transmission lines, the skin-effect losses are far greater than the dielectric losses. We include both, however, for completeness.

The lossy transmission-line model is shown in Figure 1.5. In any practical transmission line $R_l << \omega L_l$ and $G_l << \omega C_l$. This results in a line with tolerable losses, in practice, and leads to a *low-loss approximation*. This approximation makes life much easier and is entirely valid for any practical transmission line.

With this assumption, it is easy to modify the preceding equations to include losses. First, the propagation constant γ becomes

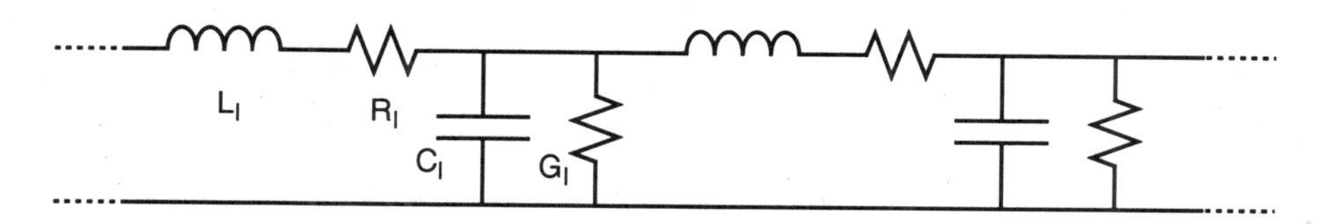

Figure 1.5 A lossy transmission line has both series resistance, R_l, and shunt conductance, G_l, which represent skin-effect losses in the conductors and various types of losses in the dielectric.

$$\gamma = \alpha + j\beta \tag{1.41}$$

With this change Equation (1.26) is still valid. Note that, for the forward-propagating wave,

$$V_f(t, z) = V_f(t, 0)\exp(-\alpha z)\exp(-j\beta z) \tag{1.42}$$

and the wave is attenuated exponentially with distance. Equations (1.23) and (1.24) can be reformulated with $R_l + j\omega L_l$ instead of $j\omega L_l$ and $G_l + j\omega C_l$ instead of $j\omega C_l$. We obtain

$$\gamma = \sqrt{(R_l + j\omega L_l)(G_l + j\omega C_l)} \tag{1.43}$$

For low-loss lines this can be manipulated into the form

$$\gamma = \frac{1}{2}\left(\frac{R_l}{Z_0} + G_l Z_0\right) + j\omega\sqrt{L_l C_l} \tag{1.44}$$

and α and β are easily extracted.

α is the loss of the line in nepers per unit of length; the loss in decibels per length is simply 8.686 α. The conventional wisdom about transmission line loss is that losses "get worse" as frequency increases. Indeed, in virtually all practical lines, where the series resistance dominates, the loss *per length* increases approximately as the square root of frequency. However, a wavelength is inversely proportional to frequency, so the loss *per wavelength* decreases as the square root of frequency. Many types of microwave components use fractional-wavelength transmission lines, and the losses of such components generally *decrease* as frequency increases.

In most practical transmission lines, the loss in G_l is much less than the loss in R_l. For this reason losses in the shunt conductance frequently are ignored.

1.2.7 Coupled Lines

A delightful variety of useful components can be made by placing two transmission lines in close proximity, so that energy from one is coupled to the other. Usually two

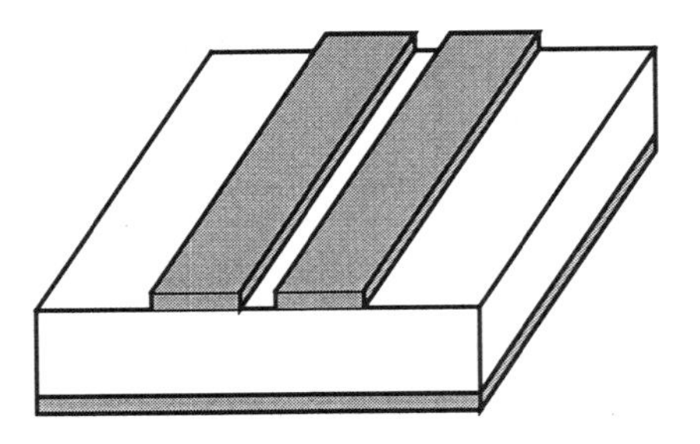

Figure 1.6 Symmetrical coupled microstrip lines.

identical symmetrical lines are used, although in some cases an asymmetrical set of multiple coupled lines can be valuable. We'll try to resist the urge to cover all possibilities and focus on the symmetrical pair only.

Figure 1.6 shows a set of coupled microstrip transmission lines (see Section 1.3.3). Analyzing the coupling can be done easily by making use of another slick trick. We first make the obvious observation that the set of lines is a linear system. As such, it obeys superposition. Because of this property, we can convert the circuit in Figure 1.7(a) to the two in Figure 1.7(b), analyze the individual circuits, and add the results. The two circuits in Figure 1.7(b) are symmetrical and thus much easier to analyze than the circuit in Figure 1.7(a) alone. They are called the *even-mode* and *odd-mode* circuits. The characteristic impedance of a wave on a single conductor of each circuit is called the *even-mode* or *odd-mode characteristic impedance*, respectively. The even-mode and odd-mode phase velocity and loss are defined similarly.

As with single transmission lines, determining the even-mode and odd-mode properties of coupled lines is a problem in electromagnetics. The general remarks in Section 1.3 regarding planar transmission lines are valid for planar coupled lines as well.

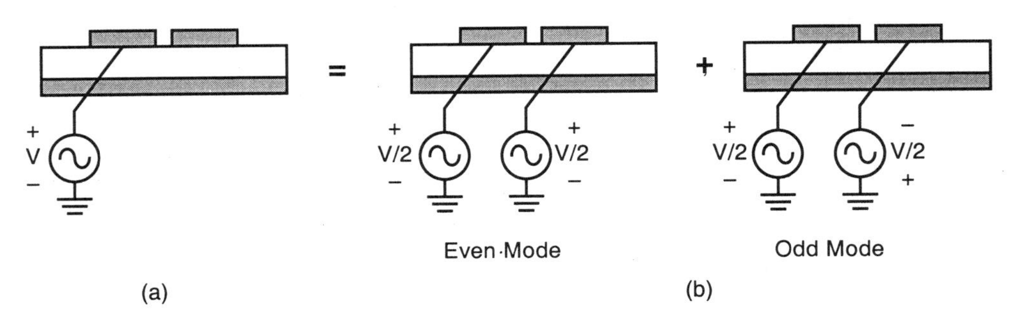

Figure 1.7 An asymmetrically excited pair of coupled lines (a) is equivalent to two sets of symmetrically excited coupled lines (b). The results of analyzing the two circuits in (b) are simply added.

Coupled lines are used for microwave hybrids and directional couplers; these are discussed in Sections 1.7.1 and 1.7.2. Coupled-line baluns are essential parts of microwave mixers, frequency multipliers, and other balanced circuits. In balun designs for mixers (Chapter 3), an appreciation of the even-mode and odd-mode characteristics of such lines is essential.

1.3 PLANAR TRANSMISSION LINES

Planar transmission lines are unbalanced transmission lines consisting of one or more flat conductors on a thin dielectric substrate with a ground surface. Such lines are the mainstay of modern RF and microwave hybrid and monolithic circuit technologies. For most circuits, some type of planar transmission medium is the only thing that makes sense.

The term *planar* is used somewhat loosely. Most planar transmission lines are not strictly planar, in the sense that they are confined to a single, flat surface. We include in this category such transmission media as stripline, suspended-substrate stripline, and microstrip. These are not really planar, but they are more or less flat. Close enough.

1.3.1 Substrate Materials

Many types of substrates are used in RF and microwave technology, including crystalline, amorphous crystalline, ceramic, and composite materials. In a monolithic circuit the substrate is an undoped semiconductor.

The dielectric material is an important part of the transmission line, as it determines many characteristics of the circuit in which it is used. Table 1.2 lists some of the most popular materials, and we discuss their merits in detail below.

Fused Silica

Loosely called *quartz*, its single-crystal form, fused silica has a number of very good and very bad properties. It is one of the few high-quality materials that have a low dielectric constant. Its dielectric constant is 3.78, much lower than other hard substrates but not as low as the composite materials. This low dielectric constant, combined with low loss and good smoothness, makes fused silica seemingly ideal for millimeter-wave circuits. Unfortunately, fused silica is also very brittle, making it difficult to handle and to fabricate, and its smoothness makes good metal adhesion difficult to obtain. Fused silica has a low thermal expansion coefficient; it is matched only to Invar or Kovar, metal alloys that are expensive and difficult to machine. If mounted on brass or aluminum, stress caused by temperature changes can crack the substrate.

Table 1.2 Substrate Materials

Material	Type of Material	Dielectric Constant	Loss Tangent	Other Characteristics
Fused Silica	Amorphous form of quartz (SiO_2)	3.78	< 0.0001 to at least 20 GHz	Expensive, brittle; difficult to obtain good metal adhesion.
Alumina	Ceramic form of alumina (Al_2O_4)	9.0 – 10.0	<0.0015 to 25 GHz	Characteristics depend on manufacture; $k = 9.8$ is most common.
Sapphire	Crystalline alumina (Al_2O_4)	8.6 horizontal, 10.55 vertical	<0.0015 in all directions	Electrically anisotropic.
RT Duroid® 5880[a]	Composite; PTFE[b]-fiberglass	2.20	0.0009 at 10 GHz	Low-cost "soft" substrate; widely used.
RT Duroid® 5870[a]	Composite; PTFE-fiberglass	2.33	0.0012 at 10 GHz	Low-cost "soft" substrate; widely used.
RT Duroid® 6006[a]	Composite; ceramic-PTFE	6.15	0.0019 at 10 GHz	Not mechanically as good as other materials
Silicon	Crystal (Si)	11.9	Very lossy	Dielectric loss is a problem for RF/MW circuits.
Gallium Arsenide	Crystal (GaAs)	12.9	Typically 0.001	Used for monolithic circuits only.
Indium Phosphide	Crystal (InP)	12.4	Typically 0.001; depends on purity	If you're using this exotic monolithic-circuit material, presumably you know something about it!

a. Rogers Corp., Chandler, Arizona
b. Polytetrafluoroethylene, or Teflon®

Metallizations on fused silica usually consist of a very thin sputtered adhesion layer with a top layer of plated gold. The adhesion layer is lossy, and its thickness strongly affects the loss of transmission lines fabricated on this substrate material.

Alumina

Alumina is one of those wonderful materials that does nothing especially well but does everything at least adequately. As such, it is one of the most frequently used materials in microwave technology.

Alumina is the ceramic form of sapphire (see below). It is a moderately expensive substrate but still the least expensive of the "hard" substrates. It is very hard, temperature-stable, and has good thermal conductivity. Although its thermal expansion coefficient is not well matched to brass or aluminum, alumina is so strong that it does not crack easily when bonded to a thermally mismatched surface, even at extreme temperatures. Alumina can be polished to high smoothness, if necessary, and metal adhesion is good. Although hard, alumina can be cut easily with a diamond substrate saw or a laser; holes can be made with a laser or a carbide tool.

Alumina has a high dielectric constant, usually 9.5 to 10.0. The precise value depends on the manufacturing process and therefore must be specified by the manufacturer. Because of this high value, millimeter-wave circuits on alumina are uncomfortably small and dispersive. For this reason alumina is not used extensively above 20 GHz.

The most common metallization is gold. A very thin adhesion layer is used between the gold and the substrate. Occasionally a barrier layer is deposited between the gold and adhesion layers to prevent chemical reactions at high processing temperatures. Circuits that require soldering often use a copper metallization with a light gold plating to prevent corrosion. Common metals for the adhesion layer are nichrome and titanium-tungsten. When nichrome is used, the adhesion layer also can be used to make thin-film resistors. A layer of tantalum nitride also can be used for resistors; this material handles high temperatures better than nichrome, but tantalum-nitride resistors have a higher temperature coefficient.

Sapphire

Sapphire is the crystalline form of aluminum oxide (Al_2O_4). It is relatively expensive. Its only advantage over alumina is its extreme smoothness, which minimizes conductor loss, and slightly lower dielectric loss. Sapphire is electrically anisotropic: its dielectric constant depends on the direction of the electric field in the material. It is 8.6 in a plane and 10.55 in the direction parallel to that plane. Sapphire usually is cut so that the $k = 8.6$ plane is parallel to the ground plane. This makes the characteristics of microstrip lines independent of their orientation, but it causes the difference between even- and odd-mode phase velocities in coupled lines to be worse than in an isotropic material.

The metallization is invariably gold with an adhesion layer.

Composite Materials

Table 1.2 lists three different composite substrates. Duroid® 5880 and 5870 are widely used. Duroid® 6006 is used somewhat less widely, but it is a good example of the main advantage of such substrates: they are available in a wide variety of dielectric constants. The three substrates listed in the table are typical; many other types of composite materials are available, from many manufacturers.

Composite materials often are called "soft substrates," because they are usually made from flexible plastics. The most common form is polytetrafluoroethylene (better known by its trade name, Teflon®), loaded with glass fibers or ceramic powder. This is both an advantage and disadvantage; the soft material is easy to handle and inexpensive to fabricate, but the mechanical and thermal properties are not as good as those of "hard" substrates. The thermal conductivity may be very low.

Composite substrates are not as consistent in their characteristics as other materials. Anyone who uses them should demand from the manufacturer firm guarantees about their characteristics. The following are some concerns:

- Tolerance of the dielectric constant;
- Variation of the dielectric constant and loss tangent with frequency and temperature;
- Electrical anisotropy;
- Thermal expansion coefficient;
- Moisture absorption;
- Volume and surface resistivity.

Composite materials almost always use copper for their conductors. Occasionally a light gold plating is applied to prevent corrosion. Strangely, the metal thickness is specified in ounces per square foot; a "1-oz." copper metallization is 1.4 mils (35 μm) thick. Typical thicknesses vary from 1/8 oz., used where fine definition is needed, to 2 oz., for high current densities. To survive flexing, metallizations on composite substrates generally are thicker than metallizations on hard substrates. Because of their thickness, they are subject to greater undercutting along the edges of conductors when etched.

Monolithic Substrates

Your choice of monolithic substrates probably will be based on cost and performance of semiconductor devices. Nonetheless, microstrip characteristics may still be factors in the choice of monolithic technologies.

The properties of gallium arsenide (GaAs) and indium phosphide (InP) are valid for related heterojunction technologies as well as simple GaAs and InP monolithic circuits. GaAs and InP have the significant advantage, compared to silicon (Si), of very low bulk conductivity. Silicon, in contrast, has such high conductivity that it is almost useless for monolithic circuits requiring microstrip structures. Silicon is used occasionally for microwave circuits where transmission-line structures are not needed. High resistivity silicon can be made, but it is almost as expensive as GaAs and performance is not as good.

1.3.2 Characteristics of Planar Transmission Lines

We examined a number of characteristics of transmission lines in Section 1.2: characteristic impedance, phase velocity, dispersion, and loss. Clearly, we need to determine those characteristics for planar transmission lines. In some cases we approach the matter a little differently from the methods in Section 1.2, more for reasons of tradition than technology. We also look into special kinds of problems and advantages presented by each type of transmission line.

How do we determine these characteristics? One of two ways: the hard way or the easy way. The hard way is to make an electromagnetic analysis of the line. Although our western, Calvinist ethic may tempt us to assume that this method must be "best," it is not necessarily optimum in any sense, sometimes not even the most accurate. For example, the first electromagnetic analyses of microstrip were based on conformal transformations, which were not accurate for very wide or very narrow lines. Later analyses, based on moment methods, required even more work but were much more accurate; the best known of these is by Bryant and Weiss [2]. The easy way is to use a set of empirical formulas (usually derived by fitting to the most accurate numerical data). Especially for the most common transmission lines, such as microstrip, the empirical formulas have been so refined that their error is often well below 1% for practical dimensions.

Characteristic Impedance

Even if you don't need to know anything else about a transmission line, you probably need to know its characteristic impedance. This is very straightforward. The only complication, in the case of microstrip, CPW, and similar "open" structures, is that housing components—both the top and the sidewalls of the circuit's enclosure—can affect the characteristic impedance. The simple solution to this problem is to keep the top and the sidewalls well away from the line and to forget about the problem. In many circuits, however, this may not be possible.

Phase Velocity

Usually, phase velocity is not explicitly calculated. Instead, the quantity of interest is the *effective dielectric constant*. The phase velocity is

$$v_p = \frac{c}{\sqrt{k_e}} \tag{1.45}$$

where k_e is the effective dielectric constant and, as before, c is the velocity of light in free space. In other words, the phase velocity has the value of a line having a homogeneous dielectric of k_e. k_e is always less than the substrate's dielectric constant. It can also be expressed as

$$k_e = \frac{C_l}{C_0} \tag{1.46}$$

where, as before, C_l is the capacitance per unit of length and C_0 is the capacitance per unit of length when the substrate's dielectric constant is 1.0.

Phase velocity, like characteristic impedance, is affected by the proximity of the housing's top and sidewalls.

Dispersion

Here's where things get sticky. All the structures in Table 1.1 except stripline, when it has a homogeneous dielectric, are inherently dispersive, and, especially at high frequencies, we must take dispersion seriously. The usual method for dealing with dispersion is to determine the line's characteristic impedance and effective dielectric constant by a quasistatic analysis and to correct for dispersion by means of a set of empirical equations. Unfortunately, the various sets of empirical dispersion equations are not nearly as accurate as those for characteristic impedance and effective dielectric constant, and people are still arguing the question of which is best.

Loss

Transmission-line loss is another sticky consideration. Again, empirical analyses of loss are reasonably accurate for microstrip and somewhat less accurate for other types of lines. The greater problem is to determine the conductivity of the line's conductors, the most important parameter in establishing a line's loss. The conductivity is always lower than the textbook values, which apply to perfect, bulk conductors measured at dc. Skin effect combined with surface roughness of the conductor decreases the apparent surface conductivity at high frequencies, and the graininess of electroplated metallizations decreases its bulk conductivity compared to the dc value. In some substrates the resistance of the adhesion layer can have a measurable effect on the loss; if the thickness of the layer is not well controlled in manufacture, the loss can be surprisingly high. Remember, the current in a microstrip line is mostly on the underside of the conductor. That's where we put the adhesion layer.

One good way to determine the conductivity is to measure the transmission-line loss and to work backward through the empirical equations to obtain the conductivity. This value then can be used to estimate the losses of lines having other dimensions.

Even when losses are accurately modeled, they can be surprisingly high. High losses can be caused by a combination of standing waves and lossy circuit elements. For example, if a lossy interconnection is located at a high-current point in the line, its loss can be increased substantially.

Other Troublesome Phenomena

A number of other phenomena cannot be quantified easily but still can cause trouble. A few of these phenomena are the following:

- *Radiation*: An open transmission line, like an antenna, can radiate. Radiation is obviously a loss mechanism but also causes coupling to other structures in the circuit, primarily the housing and the substrate, that can act as resonators. The result is spurious resonances (or, to use the technical term, *glitches*).
- *Surface waves*: Lines on low-dielectric-constant substrates radiate. On high-dielectric-constant substrates, they excite surface waves, which are guided by the electric discontinuity between the substrate and the air above it. Although radiation and surface waves are distinctly different phenomena, from the designer's standpoint there is little practical difference. Both cause the same types of problems: loss and spurious resonances.
- *Spurious modes*: Discontinuities can generate unwanted modes on the line. The substrate, acting like a dielectric resonator, can have several modes, each causing a spurious resonance at its own resonant frequency. Certain types of lines, especially CPW, are notorious for generating substrate resonances. Waveguide-like modes can be generated in the housing.

The simplest cure for these problems is to keep both the substrate and the housing as small as possible, ideally less than one-half wavelength in any dimension. If such small dimensions are impossible, keeping the housing less than one half wavelength wide may be good enough. Mounting lossy material in the housing, to absorb radiated energy, is another useful, if inelegant, technique.

1.3.3 Microstrip

The great majority of planar circuits are realized in microstrip. Microstrip is a practical medium for a wide variety of components and is a natural choice for large, integrated systems.

There is a wide variety of sets of empirical equations for microstrip. Bahl and Trivedi [3] present an excellent treatment of microstrip, including design equations for characteristic impedance, effective dielectric constant, and loss. March [4] completes and updates those equations somewhat and includes the effects of a conductive cover. March's equations are quite accurate for practical values of impedance (approximately 25 to 100 ohms), usually well within 1%. Wadell [1] includes a set of microstrip equations and treats a number of variations, as well:

microstrip with a truncated ground plane, a dielectric overlay, and other such variants. The program WINLIN, a component of the program C/NL2 [5], analyzes microstrip by both an empirical method and a numerical, quasistatic analysis similar to that of Bryant and Weiss [2]. For single lines (but not necessarily coupled lines), the difference in accuracy between the best empirical methods and numerical analyses is virtually insignificant.

Microstrip, like most planar circuits, is a "quasi-TEM" transmission line. This means that it is usually treated as a TEM line at frequencies low enough for dispersion to be negligible. At higher frequencies, dispersion corrections are usually necessary. Again, a number of methods exist. One of the most popular and most accurate is that of Kirschning and Jansen [6]. Another good one is by Wells and Pramanick [10]. Some of these methods are compared in other references [7,8].

Higher-order modes in microstrip are, of course, possible. A simple approximate expression for the cutoff frequency of the lowest non-TEM mode is

$$f_c = \frac{75}{h\sqrt{k-1}} \tag{1.47}$$

where f_c is in gigahertz and h is in millimeters. This expression implies that moding is most troublesome at high frequencies on thick, high-dielectric-constant substrates.

Empirical models for coupled microstrip lines are not as accurate as those for single lines. The best empirical model, which includes dispersion, is that of Kirschning and Jansen [9]. Their model was designed for hybrid circuits; the range of dimensions over which it is accurate may not be applicable to monolithics. In particular, the model is formulated for zero-thickness conductors, a limitation that may be troublesome in many types of monolithic coupled-line components.

Djordjevic et al. [11] have published a program that performs a quasistatic moment-method analysis of a wide variety of symmetrical and asymmetrical coupled lines. It accommodates up to 12 lines, tolerates extreme dimensions, and accounts for conductor thickness. Since monolithic circuits are very small, dispersion correction rarely is required. This program is ideal for designing coupled-line components in monolithic circuits.

1.3.4 Coplanar Waveguide

For many purposes CPW is a good alternative to microstrip. In CPW the ground surfaces are alongside the strip conductor instead of underneath it. This configuration causes many characteristics to differ from those of microstrip. First, the fields are not as fully contained in the dielectric and extend farther into the air above the substrate. This causes dispersion and radiation to be worse in CPW than in microstrip. Second, the currents are more strongly concentrated in the edges of the conductors. Because the edges are likely to be much rougher than the surfaces, losses are higher.

Nevertheless, CPW has significant advantages over microstrip for monolithic circuits. The most important is that ground connections can be made on the surface of the substrate; there is no need for "via" holes, which are used to make ground connections in microstrip circuits. CPW grounds usually have much less inductance than microstrip vias, an important consideration for many types of high-frequency circuits. Another important advantage is size. CPW conductors can be very narrow, even with low characteristic impedances. Low-impedance microstrip lines often are impractically wide. Finally, CPW is much less sensitive to substrate thickness than microstrip, so the thinning of the monolithic substrate is much less critical. CPW monolithic circuits often are not thinned at all.

The most advanced quasi-TEM treatment of CPW has been presented by Heinrich [12]. This analysis includes explicit expressions for inductance, capacitance, and resistance. From these expressions and the equations in Section 1.2, all important quantities can be calculated. Wadell [1] gives further information on CPW, including equations for related geometries.

CPW is subject to moding. The fields in CPW are especially adept at generating a parallel-plate resonator mode in the dielectric substrate. The effects of this mode are best minimized by making the dimensions of the circuit less than one-half wavelength (based on the phase velocity of a wave in the dielectric) in any dimension. If this is impossible, occasional via connections between the top and the bottom ground surfaces can be effective in removing spurious resonances.

1.3.5 Stripline

Stripline is one of the oldest types of planar transmission media, developed in the late 1950s and originally called *triplate.*[3] Of the lines listed in Table 1.1, stripline is the only true TEM transmission line. As such, it is nondispersive, but it is not immune to moding, especially if the strip conductor is not centered evenly between the ground planes.

Stripline components invariably use composite substrates. One technique is to create a sandwich of two substrates, one having a ground plane and a strip conductor, the other having only the ground plane. These two substrates are clamped firmly together to prevent the formation of an air gap, which would create variations in the dielectric constant of the medium between the ground planes.

Stripline conductors are relatively broad, making circuits larger than microstrip but relatively low loss. Because stripline uses a homogeneous dielectric, its effective dielectric constant is equal to the substrate's dielectric constant. Conformal transformation gives an accurate algebraic expression for the characteristic impedance as long as the strip conductor is negligibly thin. This can be corrected for nonzero strip thickness. Formulas and tables can be found in several of the references [1,13,14].

---

3. *Triplate* is a registered trademark of Sanders Associates, now Lockheed Sanders Co.

A factor-of-2 error that existed in one of the early papers on stripline has propagated through time and publications to the present-day literature. To check your calculations, use the conformal-transformation equations for the characteristic impedance. The characteristic impedance for the zero-thickness case is

$$Z_0 = \frac{\eta_0}{4\sqrt{k}} \frac{K(\kappa)}{K(\kappa')} \tag{1.48}$$

where

$$\kappa = \frac{1}{\cosh\left(\frac{\pi w}{2h}\right)} \qquad \kappa' = \sqrt{1-\kappa^2} \tag{1.49}$$

$K(\kappa)$ is the complete elliptic integral of the first kind, $\eta_0 = 377\ \Omega$ is the wave impedance of free space, w is the strip width, h is the spacing between ground planes, and k, as before, is the dielectric constant. Values for $K(\kappa) / K(\kappa')$ can be found in books of mathematical tables. A simple approximation for this quantity is

$$\begin{aligned} \frac{K(\kappa)}{K(\kappa')} &= \frac{1}{\pi} \ln\left(2\frac{1+\sqrt{\kappa}}{1-\sqrt{\kappa}}\right) \qquad 0.7 \le \kappa < 1.0 \\ &= \left[\frac{1}{\pi} \ln\left(2\frac{1+\sqrt{\kappa'}}{1-\sqrt{\kappa'}}\right)\right]^{-1} \qquad 0 \le \kappa \le 0.7 \end{aligned} \tag{1.50}$$

Stripline is a great medium for directional couplers. Stripline couplers can use broadside coupling to achieve high values of coupling or offset broadside coupling to achieve weaker coupling in the same structure. This is virtually impossible in microstrip or CPW, which can use only edge coupling. The homogeneous dielectric of stripline makes its even-mode and odd-mode phase velocities equal, resulting in high directivity. Broadside coupling is also possible in suspended-substrate stripline, but the mismatch between even-mode and odd-mode phase velocities, which is huge unless the dielectric constant is small, obviates its use for high-performance couplers.

Stripline is not a favored transmission medium these days, probably because it is not really suitable for components that include chip diodes, transistors, or other discrete circuit elements, and it does not integrate well with the media that do. (It is occasionally used successfully with packaged solid-state devices.) It is a good choice for many types of connectorized passive components, including filters, directional couplers, and hybrids.

The book by Howe [13] is a classic reference on stripline.

1.3.6 Suspended-Substrate Stripline

An occasional problem of stripline is that it cannot be used with hard substrates. It is easy to see why: unless the substrates are perfectly flat, an irregular air gap is left between the two substrate layers in the sandwich, and this gap has an unpredictable effect on the characteristic impedance and phase velocity. Of course, even if the substrates are perfectly flat, the metallization prevents the dielectric slabs from fitting together perfectly, and clamping them together creates stresses that can result in fracture. In many applications the mechanical properties of composite substrates (especially their high thermal expansion coefficients) make them unacceptable. Yet, except for these problems, some form of stripline may be best for the component.

What to do? One possibility is suspended-substrate stripline (SSSL). It has many of the properties of stripline but can be realized with either a hard or a soft substrate. The nonhomogeneous dielectric gives SSSL a very low effective dielectric constant, close to 1.0, and slightly lower loss than stripline. It is, however, slightly dispersive. The enclosure also is subject to waveguide-like modes, so its cross-sectional dimensions must be kept comfortably less than one-half wavelength in both width and height. An approximate expression for the lowest cutoff frequency f_c of such modes, in GHz, is

$$f_c = \frac{150}{a}\sqrt{1 - \frac{h(k-1)}{bk}} \tag{1.51}$$

where a and b are the width and the height of the channel in millimeters, h is the substrate thickness, and k is the dielectric constant.

SSSL tends to be best for high-impedance lines. Achieving a low characteristic impedance generally requires a close clearance between the conductor and the sidewall, and this creates a risk of short circuits. It also is difficult to design low-impedance lines accurately, because the sidewall usually has a notch for supporting the substrate, and most analyses of SSSL assume that the sidewall is flat.

Discontinuities can be a problem in SSSL. Because low-impedance lines must be very wide, there is a large step discontinuity between cascaded high- and low-impedance lines. (SSSL is similar to coaxial line in this regard.) Evanescent-field coupling between discontinuities in SSSL is relatively great as well, unless the channel in which the substrate is mounted is very narrow. Unfortunately, little information on modeling such discontinuities is available.

Wadell [1] gives a good treatment of SSSL. We have found Yamashita and Atsuki's variational calculation of SSSL characteristics [15] to be useful for most practical values of impedance, although the empirical equations in Wadell [1] are claimed to have ~3% accuracy, which is almost as good. Smith [16] presents an analysis of edge-coupled lines, with the charming inclusion of a Fortran listing in the appendix.

1.4 CIRCUIT ELEMENTS

Back in the dark ages, before the ascendancy of microwave monolithic technology, we used either distributed elements or lumped elements in RF and microwave circuits. Lumped elements were simple resistors, inductors, and capacitors, while distributed elements were segments of transmission lines. In general, we used lumped elements at low frequencies and distributed elements at frequencies too high for lumped ones. Everything was in its place, and everything made sense.

Today, life is not so simple. Lumped elements are used even at high frequencies and are modeled as combinations of lumped and distributed circuit elements. Even at relatively low frequencies, distributed models are sometimes used. Conversely, distributed elements (transmission-line discontinuities, for example) often are modeled by lumped elements. Everything has been turned on its head. Why?

The most important reason is the need for accuracy. Monolithic circuits cannot be tuned after manufacture, so they must be designed by a form of technological dead reckoning: design the circuit, simulate it on the computer, and expect it to work. This requires accurate models. Another reason is cost. Customers simply cannot afford to pay for an engineer to sit at a bench and "tweak" circuits for days on end. It is much less expensive to do the modeling once and to apply it to all future circuits. A good model is, in effect, a valuable investment with an immediate and high rate of return.

1.4.1 Resistors

In planar circuits a resistor is realized as a patch of resistive material deposited on the substrate or as a chip component mounted on the substrate and connected by bonded wires, solder, or conductive adhesives. In thin-film circuits the resistive material usually is made from the adhesion layer, which is exposed by etching away the gold top metal. In thick-film circuits the resistor usually is a patch of resistive ink deposited on the substrate. In monolithic circuits resistors are fabricated by deposition of metal onto the substrate. Chip resistors consist of a tiny piece of ceramic (usually alumina) with a resistive film deposited on one surface. The terminals usually are copper or nickel with a gold-plated layer. This nickel layer allows the chip to be soldered into the circuit.

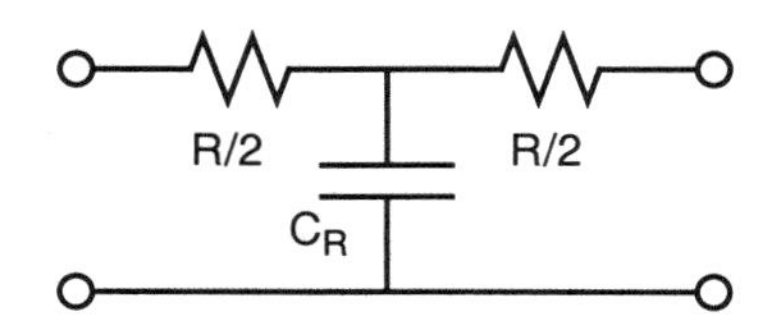

Figure 1.8 A film resistor is modeled as an RC transmission line, which is in turn modeled by lumped elements.

A thin- or thick-film resistor is essentially an RC transmission line having series resistance and shunt capacitance. If its length is much less than a wavelength, it can be modeled as shown in Figure 1.8. The capacitance can be found from microstrip-line analysis. Substituting (1.12) into (1.13) and using k_{eff} instead of k gives

$$C_R = \frac{\sqrt{k_{eff}}}{cZ_0} l \tag{1.52}$$

where l is the length of the resistor and Z_0 and k_{eff} are those of a microstrip line having the same dimensions as the resistor. The resistance of a film resistor is given in ohms per square, R_{sq}; a square resistor of any size has the same resistance. Thus,

$$R = R_{sq} \frac{l}{w} \tag{1.53}$$

where l and w are the length and width of the resistor, respectively.

The main limitation of this model is that it does not account fully for the distributed nature of the resistor. Nonetheless, it usually is adequate as long as the length is a small fraction of one wavelength. Modeling longer resistors is much more difficult. Don't be tempted to treat a long resistor as a lossy microstrip line; remember, the microstrip loss equations are a low-loss approximation, and a resistor certainly is not low loss. It generally is a bad idea to use long resistors in circuits that depend critically on the resistor's RF characteristics.

Film resistors are limited in power dissipation and current density. You are most likely to risk exceeding these limits in monolithic circuits.

1.4.2 Capacitors

A wide variety of capacitors are used in RF and microwave circuits. Chip capacitors can be as simple as a small dielectric slab, coated on both sides by metal, or as complex as multiplate structures embedded in a ceramic dielectric. Their contacts can have a variety of metallizations; gold, nickel-gold, silver, palladium-silver, and tinned nickel are the most common. Hybrid circuits usually use chip capacitors. Low-value capacitors occasionally are realized in both hybrid and monolithic circuits as interdigital capacitors; these consist of a number of short coupled-line sections in parallel. Monolithic circuits usually use metal-insulator-metal (MIM) capacitors, essentially parallel-plate devices. Because MIM and chip capacitors require different models, we view them separately.

RF Chip Capacitors

The equivalent circuit of an RF chip capacitor is shown in Figure 1.9. The inductance comes from the length of the chip, and the resistance models both the

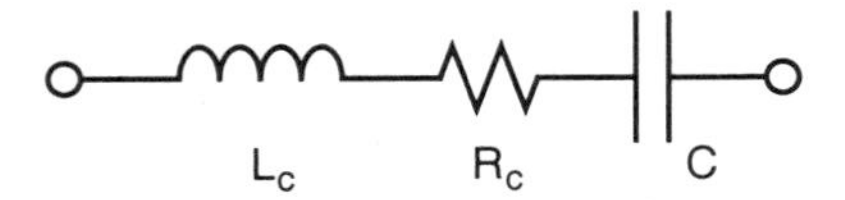

Figure 1.9 A chip capacitor is modeled as a series RLC circuit. Occasionally a shunt resistor is used to model dielectric losses, but this presents obvious problems at low frequencies.

metal and dielectric losses. The capacitance, C, is simply the low-frequency capacitance of the chip; the inductance, L_C, is found by measuring the capacitor's series resonant frequency. Since the capacitor's losses are frequency-dependent, the equivalent series resistance, R_C, should be determined from impedance measurements over a range of frequencies. Often only the capacitor's Q is specified, at some standard test frequency:

$$Q_c = \frac{X}{R_c} = \frac{1}{\omega R_c C} \tag{1.54}$$

Remember, R_C, as well as ωC, is frequency dependent, so (1.54) generally cannot be used to scale the Q in frequency.

The capacitor's series resonant frequency is an important quantity. The capacitor has its expected capacitance only at frequencies well below resonance. On the other hand, a capacitor used as a dc block is best operated at its series resonant frequency. At high frequencies some chip capacitors exhibit a parallel resonance as well, caused by the chip's inductance and the fringing capacitance between the terminals. The parallel resonant frequency can be difficult to specify, because it is affected by the way the chip is mounted.

Although ideal capacitors do not dissipate power, the equivalent series resistance of practical capacitors does indeed dissipate power. Capacitors can get quite hot when operated at high currents if the losses are not low. Capacitor heating is most likely to occur in the output stages of RF power amplifiers, where hot chip capacitors have even been known to melt solder connections.

MIM Capacitors in Monolithic Circuits

Monolithic MIM capacitors are much smaller, less lossy, and used at higher frequencies than RF chips. This dictates a different model. The model consists of an ideal capacitor and a number of microstrip transmission-line segments.

Figure 1.10 shows an MIM capacitor and its equivalent circuit. The capacitor consists of a dielectric layer deposited on a thin metal layer, which may be somewhat larger than the dielectric, and a top metal layer. The capacitance is simply

$$C = C_s A \tag{1.55}$$

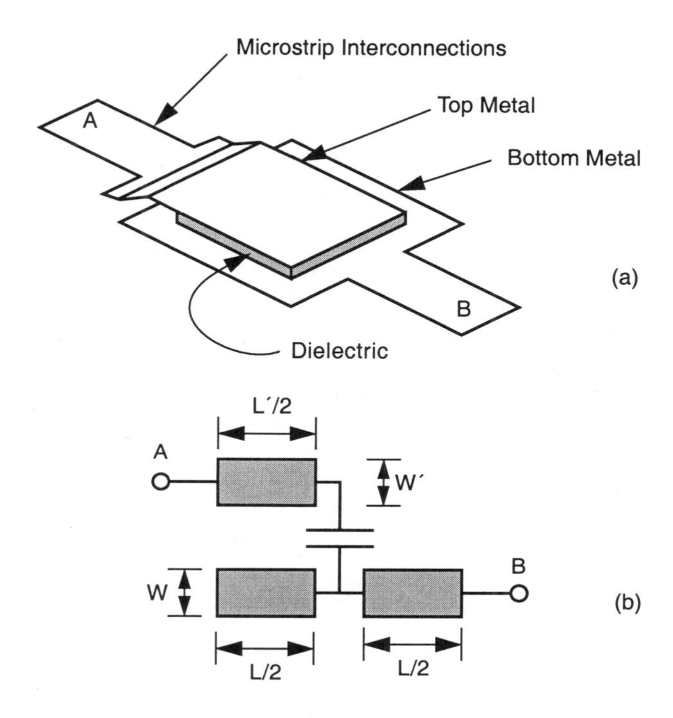

Figure 1.10 MIM capacitor (a) and its approximate model (b). *L* is the length of the lower plate and *W* is its width. *L′* and *W′* are respective quantities for the top plate.

where C_s is the sheet capacitance, a constant of the process, and A is the area of the dielectric slab. C_s is usually 150 to 300 pF/mm^2. The segments connected to the terminals A and B account for the length of the capacitor; the open stub accounts for the additional capacitance between the bottom metal and the ground plane.

We have not included a resistor in the model and thus have not accounted for losses in the dielectric or metallization. There is some justification for this. MIM capacitors used in monolithic circuits are very small, and for most purposes their losses are negligible. Losses can be included in the microstrip lines if necessary, although the losses in these conductors may be substantially different from those of a simple microstrip on the semiconductor substrate.

1.4.3 Inductors

So many types of inductors are used in RF and microwave circuits that it is virtually impossible to create a general model for them. At the lower RF frequencies wire-wound inductors are common, although as frequency increases, (say, above 400 MHz) they degenerate into a one- or two-turn loop, a "hairpin," or even a straight piece of wire. (When inductors get this small, it probably is time to start thinking about using distributed components!) As well as inductance, wire-wound inductors have capacitance between their turns. This capacitance creates a parallel resonance, and unless the inductor is operated well below this resonant frequency, the reactance of the inductor may be very different from what was expected. Unfortunately, predicting this resonant frequency is not easy, especially in view of the wide variety of shapes and sizes used in such inductors. Chip inductors, however, which are available in discrete values and sizes, can be measured, and their Qs and parallel resonant frequencies should be available from their manufacturers.

Planar spiral inductors are used in monolithic circuits and occasionally in hybrid circuits. They are a good way to achieve a high inductance (well, high by microwave standards) in a small space. Figure 1.11 shows a microstrip spiral inductor and its model. The model consists of three capacitors and a resistor as well as the inductor. The resistor models the resistive losses in the spiral. The capacitor, C_p, models the interwinding capacitance, and C_1 and C_2 model the capacitance between the spiral and ground. Determining the values of these elements is no small task. The only practical method is to measure S parameters of test inductors and to fit the element values to measured S parameters on the computer by numerical optimization. The values of these elements then can be scaled by the number of terms according to a scaling formula.

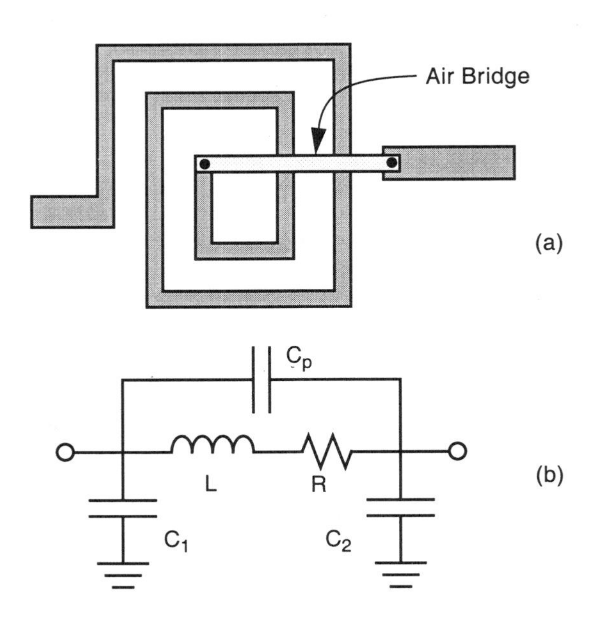

Figure 1.11 A spiral inductor (a) and its equivalent circuit (b). The resistor accounts for loss in the spiral, C_p accounts for capacitance between the windings, and C_1 and C_2 model the capacitance between the turns and the ground plane.

Spiral inductors, like wirewound inductors, exhibit a parallel resonance. This resonance establishes the upper limit of their practical frequency range: the greater the number of turns, the lower the resonant frequency.

1.5 DISTRIBUTED CIRCUIT ELEMENTS

According to the conventional wisdom, distributed circuit elements are used at high frequencies, where the parasitics of lumped elements become so great as to make lumped elements impractical. In fact, as long as the parasitics are predictable, lumped elements can be used at remarkably high frequencies, well into the millimeter-wave range. The key is *predictability*, and this is what we obtain from well-conceived models. Even so, distributed elements—essentially, transmission-line segments—usually can be characterized more accurately than lumped elements.

Distributed elements may be too big for many types of RF circuits, especially monolithics. Distributed circuit elements must be an appreciable fraction of a wavelength long; at 1 GHz, this is at least a few centimeters. This requirement presents a dilemma for the designer, especially for circuits between 1 and 5 GHz. In this frequency range, the sizes of distributed elements may make them impractical, but the parasitics of lumped elements in chip form may be too great.

Table 1.3 summarizes the characteristics of the most important distributed elements used in RF and microwave circuits. These elements can approximate inductors, capacitors, or resonators. We emphasize the word *approximate*; although these elements do indeed exhibit inductive or capacitive reactance, their reactance does not vary with frequency in the same way as a lumped inductor or capacitor.

1.5.1 Transmission-Line Stubs

Stubs are the most important of the elements in Table 1.3. A *stub* is a length of straight transmission line that is short- or open-circuited at one end and connected to a circuit at the opposite end. Stubs can approximate inductors, capacitors, or resonators. High- or low-impedance series lines also approximate series inductors or shunt capacitors, respectively, but not as accurately. Stubs are used almost exclusively as shunt elements. Although they could, in theory, be used to realize series elements, there are a couple of problems in doing so. First, the stub would have to be realized by a parallel-coupled line. The even mode on such a line would introduce shunt capacitance, so the stub would not be a series element. Second, such structures often are difficult to realize both mechanically and electrically. Usually they just don't work.

The impedance of such a stub is easy to determine. We use (1.36) with $Z_L = 0$ or $Z_L \to \infty$, for a short- or an open-circuit stub, respectively. This gives, for the short-circuit stub,

$$Z_{in} = jZ_0 \tan(\beta l) \tag{1.56}$$

Table 1.3 Distributed Circuit Elements

Structure	Characteristics and Uses	Equation	Exact or Approximate
Short-circuit stub	Inductive when $\beta l < \pi/2$	$X = Z_0 \tan(\beta l)$	Exact
Open-circuit stub	Capacitive when $\beta l < \pi/2$	$B = Y_0 \tan(\beta l)$	Exact
Quarter-wave, open-circuit stub	Equivalent to a series LC resonator	$L = \frac{\pi Z_0}{4\omega_0}$ $C = \frac{1}{\omega_0^2 L}$	Approximate; based on equating $dX / d\omega$ at resonance
Quarter-wave, short-circuit stub	Equivalent to a parallel LC resonator	$C = \frac{\pi Y_0}{4\omega_0}$ $L = \frac{1}{\omega_0^2 C}$	Approximate; based on equating $dB / d\omega$ at resonance
Radial stub	Used almost exclusively to realize a broadband short circuit	(No simple equation describes this element adequately)	No exact expression exists
High-impedance series line	Equivalent to a series inductance	$X = Z_0 \tan(\beta l)$	A fairly crude approximation; OK when $\beta l << \pi/4$
Low-impedance series line	Equivalent to a shunt capacitance	$B = Y_0 \tan(\beta l)$	A fairly crude approximation; OK when $\beta l << \pi/4$

and for the open-circuit stub,

$$Z_{in} = -jZ_0 \cot(\beta l) \tag{1.57}$$

These relations are easily manipulated into the forms shown in the table, which are purposely designed to illustrate their duality.

When a stub is an integral multiple of one-quarter wavelength long, it behaves as a resonator. The quarter-wave short-circuit stub operates as a parallel *LC* resonator and the open-circuit as a series *LC* resonator. The two are not exact equivalents. The equations for *L* and *C* of an "equivalent" *LC* resonator given in the table are derived by equating the slope parameters of the resonators [14]. The slope parameter of a series resonator is

$$x = \frac{\omega_0}{2}\frac{dX}{d\omega} \tag{1.58}$$

For a parallel resonator the slope parameter is

$$b = \frac{\omega_0}{2}\frac{dB}{d\omega} \tag{1.59}$$

where ω_0 is the resonant frequency. A little experimentation shows that the slope parameter of a series *LC* resonator is simply $\omega_0 L$ and a quarter-wave, open-circuit stub is $\pi Z_0/4$.[4] Equating these gives the entries in the table. The equivalences are accurate over approximately a 20% bandwidth.

1.5.2 Radial Stub

Figure 1.12 shows a microstrip radial stub. A radial stub is an open-circuit stub realized in radial transmission line instead of straight transmission line. It is a very useful element, primarily for providing a clean (no spurious resonances) broadband short circuit, much broader than a simple open-circuit stub. It is especially useful on bias lines in high-frequency amplifiers and similar components. Unfortunately, no simple theory describes this a radial stub adequately. Atwater [17] approximates it as a cascade of short sections, a treatment that provides a good trade-off between simplicity and accuracy. March [18] and Giannini [19] give other useful models.

Radial stubs are used almost exclusively in microstrip circuits; they could be used in stripline as well. Although radial stubs are shorter than uniform stubs, they cannot be folded or bent; therefore they take up a lot of substrate area. For this reason radial stubs are used primarily at high frequencies, where they are relatively small.

---

4. Just remember that the reactance of the equivalent inductor in the series *LC* resonator should equal the characteristic impedance of the line. This is close enough.

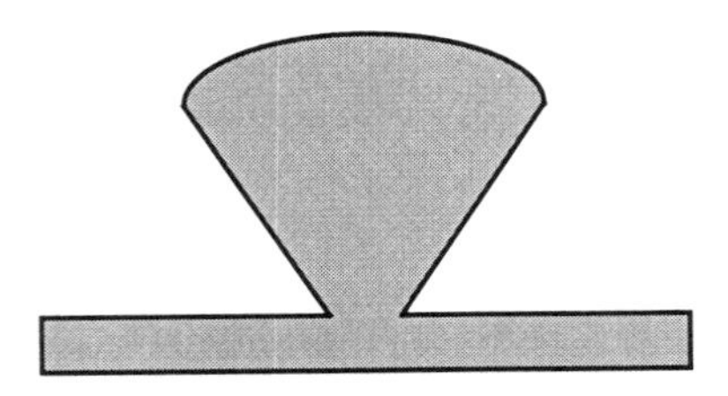

Figure 1.12 Microstrip radial stub connected in shunt with a microstrip line.

1.5.3 Series Lines

The expressions for the series lines will not be derived here. We will merely repeat the point that they are valid when $\beta l << \pi/4$, and under these conditions $\tan(\beta l) \approx \beta l$. We should also quantify what we mean by high and low impedances: we mean that they are high or low compared to the impedances locally in the circuit. For example, a filter designed for 50Ω terminations requires $Z_0 >> 50\Omega$ or $Z_0 << 50\Omega$.

In all candor, series lines do not provide very good approximations of shunt capacitors or series inductors unless the capacitance or inductance is fairly low. Even then, the discontinuities introduced by cascading low- and high-impedance sections, as would exist in a low-pass filter, for example, can be difficult to characterize accurately.

1.5.4 Discontinuities

Once we start using transmission lines to approximate circuit elements, we collide headlong with a fundamental truth: the lines must be interconnected, and each interconnection introduces discontinuities. Typical discontinuities are microstrip tee junctions, crosses, and steps in width. Especially at high frequencies, the effects of these discontinuities simply cannot be ignored.

For example, consider the microstrip step junction in Figure 1.13. The dominant effects are the inductance, caused by current crowding at the junction, and the capacitance, caused by the fringing electric field. These phenomena are modeled by the equivalent circuit in Figure 1.13 (b). The values of the inductance and capacitance depend on the dimensions of the microstrip lines and on frequency. Determining expressions for these values has been a wonderful source of employment for electromagneticists; for some examples, see Wadell [1].

Although transmission-line discontinuities can be difficult to model, we don't need very many of them. We can make a lot of nice circuits with only a microstrip tee, step, cross, bend, and open circuit. Models for these discontinuities included in circuit simulators usually are adequate for most purposes, but more extreme cases, such as a very large step in line width, might be outside the range over which they are accurate. Be careful!

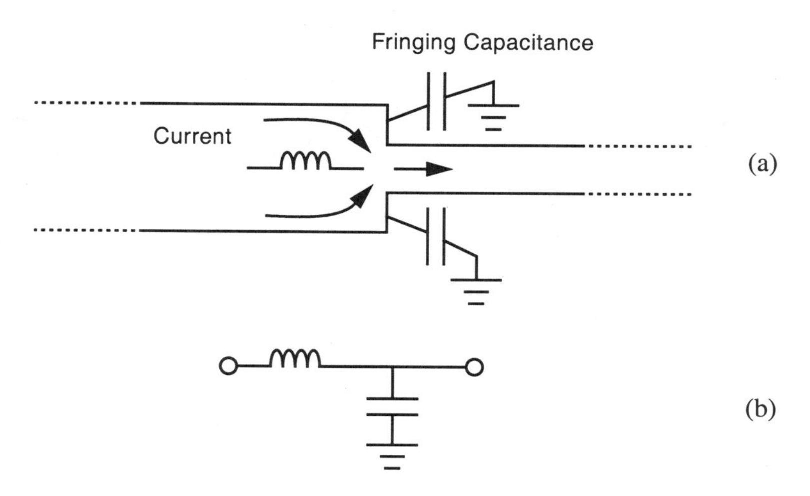

Figure 1.13 The dominant parasitics in a microstrip step junction are the fringing capacitance at the step and the inductance caused by current crowding at the junction. (a) Illustration of the source of the parasitics; (b) equivalent circuit of the step junction.

1.6 SCATTERING PARAMETERS

1.6.1 Wave Variables

It seems logical that, in circuits where traveling waves are easy to define and voltages and currents difficult, some type of wave representation would be the best characterization for a multiport. Certainly, wave characteristics are easier to measure in real circuits than voltages and currents; after all, have you ever seen a 60-GHz ammeter? Even though RF and microwave measurements must use components that are sensitive to waves (directional couplers, for example), the circuits themselves can be described in terms of either port voltages and currents or waves incident on and reflected from those ports.

The distinction between wave variables and voltage/current variables is not as great as it might appear. Sometimes we must speak of voltages and currents in places where they are difficult to define precisely; for example, in a waveguide component. At the same time, wave variables can be defined in lumped, nondistributed circuits. Fortunately, we can develop an equivalence between wave variables and terminal voltages and currents, so either can be used for analysis.

We start by considering a simple one-port circuit, with source impedance R and input impedance Z. This circuit, shown in Figure 1.14(a) can be converted into the one in Figure 1.14 (b), where the input impedance Z has been replaced by a resistor of resistance R and an impedance $Z - R$. Finally, in Figure 1.14 (c) we replace the impedance $Z - R$ by a controlled voltage source. To do this, we note that

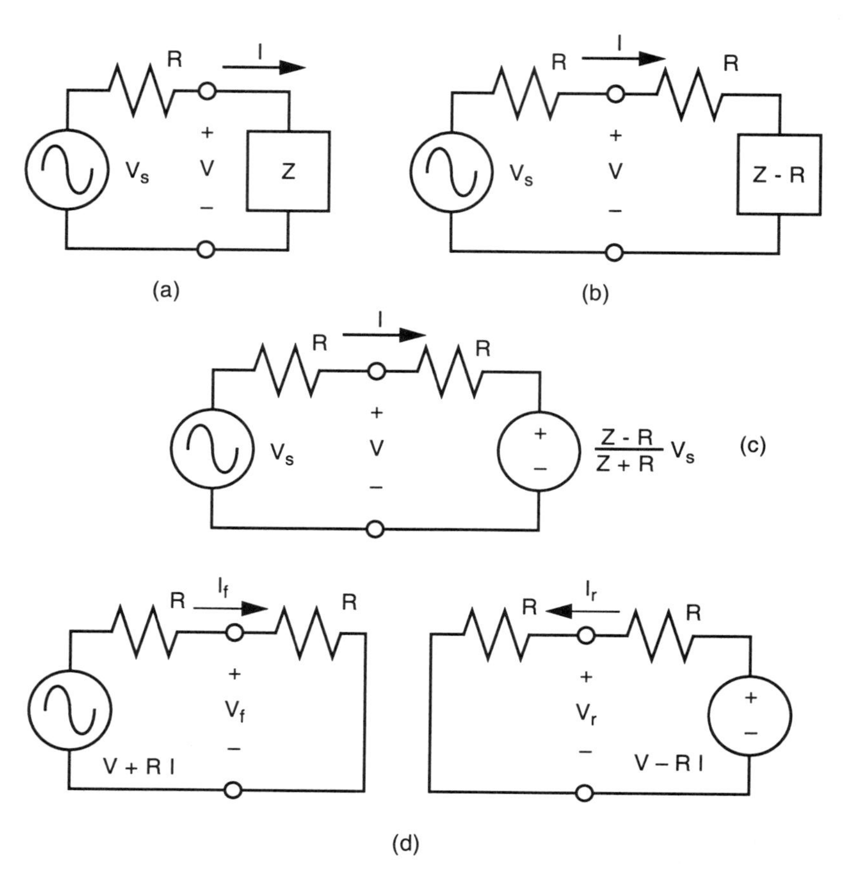

Figure 1.14 Derivation of forward- and reverse-propagating waves at a single port. The two circuits in (d) are equivalent to the original one-port (a).

$$I = \frac{V_s}{Z+R} \tag{1.60}$$

and

$$(Z-R)I = V_x = \frac{Z-R}{Z+R} V_s \tag{1.61}$$

where V_x is the controlled-source voltage and the other quantities are defined in Figure 1.14. Equation (1.61) looks a lot like (1.32), the equation for reflection coefficient. To make this more explicit, we can use superposition to convert the circuit of Figure 1.14(c) to the pair of circuits in Figure 1.14(d). The terminal voltage, V, is

$$V = V_f + V_r \tag{1.62}$$

and the current, I, is

$$I = I_f - I_r \tag{1.63}$$

Now we want to express the excitation and controlled source voltages in terms of the terminal voltage. This can be done by inspection:

$$\begin{aligned} V_s &= V + RI \\ V_x &= V - RI \end{aligned} \tag{1.64}$$

Similarly, from the circuits we can see that

$$\begin{aligned} V_f &= RI_f \\ V_r &= RI_r \end{aligned} \tag{1.65}$$

A little algebraic manipulation of (1.62) through (1.65) gives

$$\begin{aligned} V_f &= \frac{1}{2}(V + RI) \qquad & I_f &= \frac{V + RI}{2R} \\ V_r &= \frac{1}{2}(V - RI) \qquad & I_r &= \frac{V - RI}{2R} \end{aligned} \tag{1.66}$$

We now have fundamental relationships between the wave variables V_f, V_r, I_f, and I_r and the port voltages and currents. By manipulating them further, we can develop relationships between multiport parameters (for example, admittance parameters or impedance parameters) and similar multiport parameters defined in terms of wave variables.

We could stop at this point, but, like good microwave engineers, we wish (1) to normalize all quantities, so they don't depend on the resistance R, and (2) to put the wave variables into a form that expresses power, not voltage or current. Fortunately, this is easy to do. We define two new wave variables, a and b, as follows:

$$\begin{aligned} a &= \sqrt{V_f I_f} = V_f \sqrt{R}^{-1} = I_f \sqrt{R} \\ b &= \sqrt{V_r I_r} = V_r \sqrt{R}^{-1} = I_r \sqrt{R} \end{aligned} \tag{1.67}$$

Here we have the best of both worlds. We can square a or b and obtain a value that is proportional to power but also treat each of them as normalized voltages or currents, where R is the *normalization resistance*. For reasons that will become clear in a moment, we call these *scattering variables*. Finally, substituting (1.66) into (1.67) gives us a relationship between a and b and the terminal voltage and current:

$$
\begin{aligned}
a &= \frac{1}{2}\left(\frac{V}{\sqrt{R}} + I\sqrt{R}\right) \\
b &= \frac{1}{2}\left(\frac{V}{\sqrt{R}} - I\sqrt{R}\right)
\end{aligned}
\tag{1.68}
$$

From (1.67) it is easy to show that the relationship between a and b is just the reflection coefficient:

$$\frac{b}{a} = \frac{V_r}{V_f} = \frac{I_r}{I_f} = \Gamma \tag{1.69}$$

and when $Z = R = V / I$,

$$\frac{1}{2}|a|^2 = \frac{|V|^2}{2R} \tag{1.70}$$

which is the available power of the source. The analogous quantity $|b|^2 / 2$ also has units of power, of course; it represents the power in the reflected wave. The power delivered to the load, P_{del}, is just the difference between these two quantities:

$$P_{del} = \frac{1}{2}(|a|^2 - |b|^2) = P_{av}(1 - |\Gamma|^2) \tag{1.71}$$

This is a very useful relationship.

1.6.2 Traveling Waves

From the similarity between (1.61) through (1.67) and (1.30) through (1.38), we can interpret a and b as traveling waves on a transmission line, of characteristic impedance R, connected to the port. The wave a is the incident wave on the port and b is the reflected wave. These quantities are complex, so we must define a reference position for their phases. Usually this position is the plane of the port.

We use the term *reflected wave* somewhat loosely. More precisely, it is the wave traveling away from the port; it need not be a reflection in the sense used in Section 1.2.3. For example, if the component contained an independent source, b would be the outward-propagating wave generated by that source.

1.6.3 Multiport Scattering Variables

The concept we have just developed can be extended easily to multiports. In the single-port case we had

$$b = \Gamma a \tag{1.72}$$

and for multiports we have

$$\underline{b} = \underset{\sim}{S}\underline{a} \tag{1.73}$$

where the underline signifies a vector, and the tilde indicates a matrix. $\underset{\sim}{S}$ is called a *scattering matrix*, or a set of *S parameters*:

$$\begin{bmatrix} b_1 \\ b_2 \\ \cdots \\ b_N \end{bmatrix} = \begin{bmatrix} S_{11} & S_{12} & \cdots & S_{1N} \\ S_{21} & S_{22} & \cdots & S_{2N} \\ \cdots & \cdots & \cdots & \cdots \\ S_{N1} & S_{N2} & \cdots & S_{NN} \end{bmatrix} \begin{bmatrix} a_1 \\ a_2 \\ \cdots \\ a_N \end{bmatrix} \tag{1.74}$$

From what we have seen so far, it is easy to identify some of the characteristics of this matrix:

- S_{ii} is the input reflection coefficient at port i when all the other ports are terminated in their normalizing resistances (which, by the way, need not be identical).
- $|S_{ij}|^2$ is the *transducer gain* between an input at port j and an output at port i, again when all the other ports are terminated in their normalizing resistances.

The first point is obvious: $S_{ii} = b_i / a_i$ when incident waves at all other ports, a_n, $n \neq i$, are zero. This means that all other ports are terminated in their normalizing resistances and are not excited. The multiport is, in effect, reduced to a one-port and (1.72) applies.

The second point requires a little more explanation. As we claimed earlier, $|b_i|^2$ is proportional to the power in the reflected wave (to be specific, it's twice that power). Since the line is terminated in its characteristic impedance, all that power is delivered to the load at port i. Similarly, $|a_j|^2$ is twice the available power at port j. Therefore,

$$|S_{ij}|^2 = \frac{|b_i|^2}{|a_j|^2} = \frac{P_{del}}{P_{av}} \tag{1.75}$$

where P_{del} is the power delivered to the load and P_{av} is the available power of the source. The ratio of these quantities is the definition of the *transducer gain*.

1.6.4 Conversions Between Scattering Parameters and Other Parameter Sets

In a manner analogous to (1.68), we can say, for the matrices and vectors,

$$\begin{aligned} \underline{a} + \underline{b} &= \underset{\sim}{R}^{-1/2}\underline{V} \\ \underline{a} - \underline{b} &= \underset{\sim}{R}^{1/2}\underline{I} \end{aligned} \tag{1.76}$$

where $\underline{V}$ and $\underline{I}$ are vectors of port voltage and current, and $\underset{\sim}{R}^{1/2}$ is a diagonal matrix whose elements are the square roots of the normalizing resistances at each port. $\underset{\sim}{R}^{-1/2}$ is the inverse of $\underset{\sim}{R}^{1/2}$.

Equation (1.76) is the key to converting between scattering parameters and other parameter sets. This conversion simply involves (1.76) and matrix manipulations. For example, let's convert the S matrix to an impedance (Z) matrix:

$$\underline{V} = \underset{\sim}{Z}\underline{I} \tag{1.77}$$

which can be written

$$\underset{\sim}{R}^{-1/2}\underline{V} = \underset{\sim}{R}^{-1/2}\underset{\sim}{Z}\underset{\sim}{R}^{-1/2}\underset{\sim}{R}^{1/2}\underline{I} \tag{1.78}$$

We define $\underset{\sim}{Z}_n = \underset{\sim}{R}^{-1/2}\underset{\sim}{Z}\underset{\sim}{R}^{-1/2}$, which we call the *normalized impedance matrix*, and substitute (1.76):

$$\underline{a} + \underline{b} = \underset{\sim}{Z}_n(\underline{a} - \underline{b}) \tag{1.79}$$

After a little manipulation we obtain

$$\underline{b} = (\underset{\sim}{Z}_n + \underset{\sim}{1})^{-1}(\underset{\sim}{Z}_n - \underset{\sim}{1})\underline{a} \tag{1.80}$$

or

$$\underset{\sim}{S} = (\underset{\sim}{Z}_n + \underset{\sim}{1})^{-1}(\underset{\sim}{Z}_n - \underset{\sim}{1}) \tag{1.81}$$

This looks a lot like (1.32), doesn't it?

Now, if you think this is a lot of fun, try to derive a few of these:

$$\underset{\sim}{S} = \underset{\sim}{R}^{1/2}(\underset{\sim}{Z} + \underset{\sim}{R})^{-1}\underset{\sim}{R}^{1/2} \tag{1.82}$$

$$\underset{\sim}{Z} = \underset{\sim}{R}^{1/2}(2(\underset{\sim}{1} - \underset{\sim}{S})^{-1} - \underset{\sim}{1})\underset{\sim}{R}^{1/2} \tag{1.83}$$

$$\underset{\sim}{Z} = \underset{\sim}{R}^{1/2}(\underset{\sim}{1} - \underset{\sim}{S})^{-1}(\underset{\sim}{1} + \underset{\sim}{S})\underset{\sim}{R}^{1/2} \tag{1.84}$$

and, for admittance matrices,

$$\underset{\sim}{S} = (\underset{\sim}{1} + \underset{\sim}{Y}_n)^{-1}(\underset{\sim}{1} - \underset{\sim}{Y}_n) \tag{1.85}$$

$$\underset{\sim}{Y} = \underset{\sim}{R}^{-1/2}(\underset{\sim}{1} + \underset{\sim}{S})^{-1}(\underset{\sim}{1} - \underset{\sim}{S})\underset{\sim}{R}^{-1/2} \tag{1.86}$$

where $\underset{\sim}{Y}_n$, the normalized admittance matrix, is the inverse of $\underset{\sim}{Z}_n$ and $\underset{\sim}{1}$ is the identity matrix. For a complete table of conversions between two-port matrices, see Reference [20].

1.6.5 Useful Expressions for Two-Ports

Most of the components we encounter are two-ports. Even devices that are not strictly two-ports, such as transistors, often are characterized by two-port S parameters. Therefore, it is valuable to have a set of expressions that tell us the things we most want to know about two-port components.

Consider the S parameters of a two-port:

$$\begin{bmatrix} b_1 \\ b_2 \end{bmatrix} = \begin{bmatrix} S_{11} & S_{12} \\ S_{21} & S_{22} \end{bmatrix} \begin{bmatrix} a_1 \\ a_2 \end{bmatrix} \tag{1.87}$$

Suppose the output is terminated in a resistance other than the normalizing resistance. The termination has a reflection coefficient Γ_L, so

$$a_2 = \Gamma_L b_2 \tag{1.88}$$

Substituting (1.88) into (1.87) gives, with a little algebra,

$$\frac{b_2}{a_1} = \Gamma_{in} = S_{11} + \frac{S_{12}S_{21}\Gamma_L}{1 - S_{22}\Gamma_L} \tag{1.89}$$

where Γ_{in} is the input reflection coefficient. Similarly, the output reflection coefficient, Γ_{out}, is

$$\Gamma_{out} = S_{22} + \frac{S_{12}S_{21}\Gamma_S}{1 - S_{11}\Gamma_S} \tag{1.90}$$

where Γ_S is the source reflection coefficient.

An expression for the transducer gain is a little more difficult to derive. It is

$$G_T = \frac{|S_{21}|^2(1 - |\Gamma_S|^2)(1 - |\Gamma_L|^2)}{|(1 - S_{11}\Gamma_S)(1 - S_{22}\Gamma_L) - S_{12}S_{21}\Gamma_S\Gamma_L|} \tag{1.91}$$

and we can see immediately that, when $\Gamma_S = \Gamma_L = 0$, $G_T = |S_{21}|^2$, as expected.

1.7 MICROWAVE COMPONENTS

We need to be aware of a number of passive microwave components: baluns, hybrid couplers (usually just called *hybrids)*, directional couplers, and isolators. These components are used in a wide variety of system applications and as parts of other components, both active and passive.

1.7.1 Hybrid Couplers and Baluns

Hybrids

A hybrid coupler is a lossless reciprocal four-port microwave component having a specific set of properties. There are two types: 90-degree and 180-degree hybrids. The properties of an ideal hybrid are as follows:

1. All four ports are matched, in the sense that the ports' input impedances are equal to their normalizing impedances.
2. When any port is excited, the output power is divided equally between two other ports. The fourth port is isolated.
3. In a 180-degree hybrid, depending on the port chosen for the input, the outputs are either in-phase or differ in phase by 180 degrees.
4. In a 90-degree hybrid, the two outputs always differ in phase by 90 degrees, regardless of the choice of the input port.

An ideal 180-degree hybrid has the S matrix

$$S_{180} = \frac{1}{\sqrt{2}}\begin{bmatrix} 0 & 0 & 1 & 1 \\ 0 & 0 & 1 & -1 \\ 1 & 1 & 0 & 0 \\ 1 & -1 & 0 & 0 \end{bmatrix} \tag{1.92}$$

The S matrix of a 90-degree hybrid is

$$S_{90} = \frac{1}{\sqrt{2}}\begin{bmatrix} 0 & 0 & -j & 1 \\ 0 & 0 & 1 & -j \\ -j & 1 & 0 & 0 \\ 1 & -j & 0 & 0 \end{bmatrix} \tag{1.93}$$

Hybrids have a variety of uses. This book explores their applications in balanced structures, especially balanced mixers and frequency multipliers.

Ninety-degree hybrids have a surprising and useful property: if the loads connected to the output ports have equal reflection coefficients, the input reflection coefficient is always zero. (If you have a free afternoon, you can prove this to yourself. The proof is similar to the derivation of (1.89).) This property is frequently exploited to create *balanced* or, more correctly, *quadrature-coupled* amplifiers, shown in Figure 1.15. In this circuit, the input power is split by a 90-degree hybrid

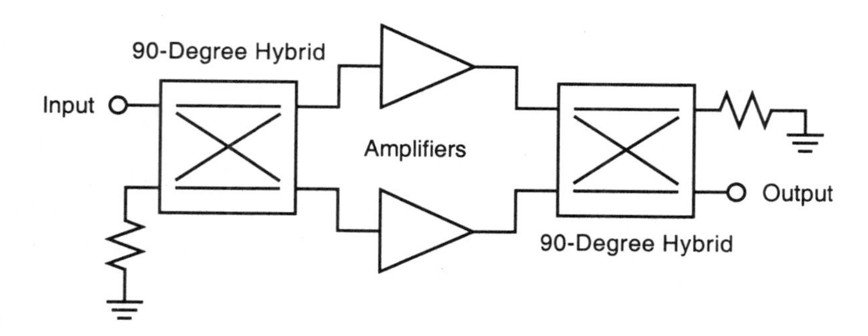

Figure 1.15 Quadrature-coupled amplifiers. By combining two amplifiers in this way, the circuit has a very low input reflection coefficient over a broad range of frequencies. Other types of components can be quadrature-coupled as well.

and applied to the inputs of two identical amplifiers. The output power is combined in a similar manner. This amplifier has the same gain as a single amplifier and the same noise figure, minus the small losses in the hybrids, of course. However, because there are now two amplifiers, the combination has twice the power-handling capability. The input and output VSWRs are very low. In theory, the input and output are perfectly matched, but imperfections in the hybrids and discontinuities in the circuits limit the input return loss to 15 to 20 dB at best.

This circuit is practical because of another interesting property of coupled-line 90-degree hybrids: although the bandwidth of the power split is limited to about an octave, the phase difference between the outputs is independent of frequency. As a result, such amplifiers frequently have bandwidths of one or two octaves, even more if some degradation from the unequal power split at the band edges can be tolerated.

Although it is most frequently applied to amplifiers, this technique can improve the input VSWR of almost any two-port. The main disadvantage is the obvious one: two components are required, but most performance characteristics are no better than a single component.

Baluns

Balun is a contraction of the words *balanced* and *unbalanced.* A balun is simply a transducer between a balanced transmission structure, such as a coaxial line, and an unbalanced structure, such as a parallel-wire line. Baluns are used most often to connect an unbalanced transmission line to a component that requires balanced excitation. We use them often in balanced mixers (Chapter 3).

It is important to distinguish between a balun and a 180-degree hybrid. A 180-degree hybrid can be used as a balun, but a balun is not a hybrid. We examine this point, along with other useful balun structures, in Section 3.1.4.

Types of Hybrids

In this book we examine only a few types of baluns and hybrids. Examples of other types can be found in References [14] and [21].

The hybrids we use in the circuits in later chapters are pretty simple: the 180-degree rat-race hybrid and the 90-degree branch-line hybrid. The design of these hybrids is described in Sections 3.2 and 3.3, respectively. Sections 3.4 and 3.5 describe two types of very useful baluns, the parallel-strip coupled-line balun and a "horseshoe" section that enhances its performance. Section 3.6 describes the star mixer and its all-important Marchand balun. Finally, the Lange coupler, perhaps the most important type of 90-degree hybrid, is described in Section 1.7.2, below.

1.7.2 Directional Couplers

Suppose we design a set of coupled lines (see Section 1.2.7 and Figure 1.6) so that (1) they are one-quarter wavelength long, and (2) the even-mode and odd-mode characteristic impedances satisfy the following relations:

$$\begin{aligned} Z_{0e} &= R\sqrt{\frac{1+k}{1-k}} \\ Z_{0o} &= R\sqrt{\frac{1-k}{1+k}} \\ R &= \sqrt{Z_{0e}Z_{0o}} \end{aligned} \tag{1.94}$$

where k is a constant between 0 and 1.0, called the *coupling coefficient*, and R is the port-normalizing impedance. We find that the structure has the S matrix,

$$S = \begin{bmatrix} 0 & 0 & -j\sqrt{1-k^2} & k \\ 0 & 0 & k & -j\sqrt{1-k^2} \\ -j\sqrt{1-k^2} & k & 0 & 0 \\ k & -j\sqrt{1-k^2} & 0 & 0 \end{bmatrix} \tag{1.95}$$

This looks a lot like the S matrix of a 90-degree hybrid, but the power split is unequal; the power coupling to one port is k^2 and to the other $1 - k^2$.

Figure 1.16 shows the coupled lines and describes the conventional nomenclature for the ports. Most interesting is the fact that the coupled port is on the same side of the coupler as the input line. This implies that the coupled wave travels in the opposite direction as the excitation wave; for this reason, these sometimes are called *backward-wave couplers*. There is no simple, intuitive explanation for this phenomenon.

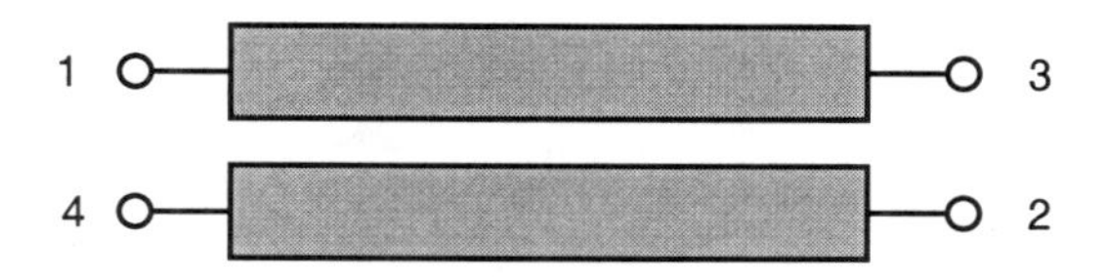

Figure 1.16 Coupled-line directional coupler, with port numbering that corresponds to Equation (1.95). Port 1 is the input port, port 4 is the coupled port, port 3 is the "through" port, and port 2 is isolated.

These couplers are very broadband. The coupling is

$$S_{41} = \frac{jk\sin(\theta)}{\sqrt{1-k^2}\cos(\theta) + j\sin(\theta)} \tag{1.96}$$

and the through-port response is

$$S_{31} = \frac{\sqrt{1-k^2}}{\sqrt{1-k^2}\cos(\theta) + j\sin(\theta)} \tag{1.97}$$

where θ is the electrical length of the coupler,

$$\theta = 2\pi\frac{l}{\lambda} \tag{1.98}$$

where l is the length of the coupled lines and λ is the wavelength. These expressions are based on an assumption that the phase velocities of the even and odd modes are the same. If they are not, as is often the case, (1.96) and (1.97) may lose accuracy. More importantly, the coupler will not work well if the phase velocities are very different. This is especially true when the coupling is weak, below 15 – 20 dB.

Finally, we define two more coupler parameters: isolation and directivity. *Isolation* is simply $|S_{21}|^2$; it is the ratio of output power at the isolated port to available input power. *Directivity* is the ratio of power at the isolated port to power at the coupled port; thus, it is $|S_{21}|^2 / |S_{41}|^2$.

90-Degree Hybrid Couplers Revisited

A comparison of (1.95) and (1.93) shows that a 90-degree hybrid is just a directional coupler with 3-dB coupling, or $k = 0.707$. Unfortunately, in edge-coupled microstrip, it is impossible to get enough coupling with a single pair of strips to achieve the necessary even- and odd-mode impedances ($Z_{0e} = 120.7\Omega$ and $Z_{0o} = 20.7\Omega$ when $R = 50\ \Omega$). The solution, originally suggested by Julius Lange [22], is to split the two strips into four and to connect alternating strips in parallel. This increases the coupling enough to make a 3-dB coupler practical. Finally, to put the outputs on the

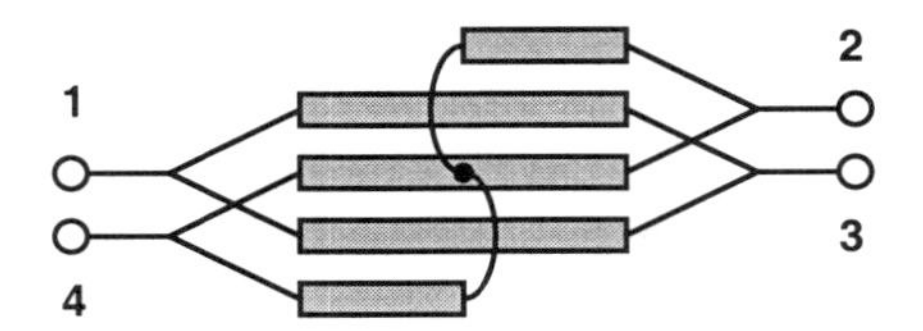

Figure 1.17 A Lange coupler is just a modified two-strip coupled-line directional coupler, similar to the one shown in Figure 1.16. The strips are split, rearranged and connected as shown to increase coupling and to put both outputs on the same side of the structure.

same side of the coupler, one outer strip is cut and moved to the opposite side. The resulting structure is called a *Lange coupler.*

For no particularly good reason, Lange couplers are not used very often in mixers and other nonlinear circuits; branch-line hybrids, discussed in Chapter 3, are more common. Lange couplers are always used in quadrature-coupled amplifiers (Figure 1.15).

1.7.3 Circulators and Isolators

The symbol for a circulator, shown in Figure 1.18, tells almost the whole story. A wave incident on port 1 emerges from port 2, a wave incident on port 2 emerges from port 3, and so on. Circulators usually are three-port components, but by interconnecting several of them, multiport circulators can be made.

The most common type of isolator is simply a circulator with a terminated port, which makes it a two-port. A signal incident on port 1 emerges from port 2, but a signal incident on port 2 disappears into the termination on what had been port 3, never to be heard from again. No matter how bad the VSWR of the port 2 termination, the input VSWR at port 1 is always unity.

A circulator is a passive, nonreciprocal component. Passive structures realized from lumped elements or ordinary materials are always reciprocal, so to realize a circulator we must use a nonreciprocal material. Circulators use a resonator made

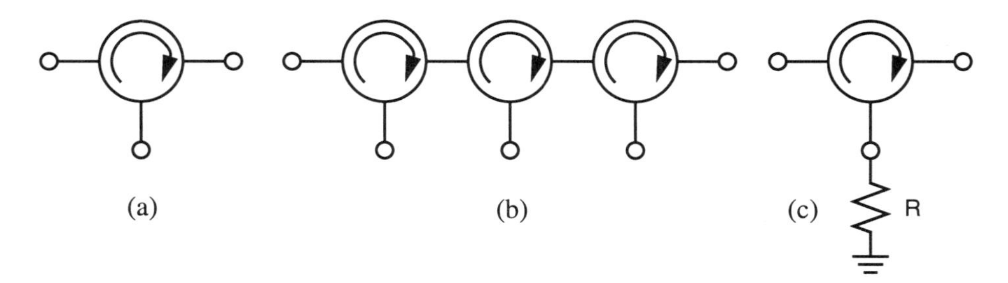

Figure 1.18 (a) Three-port circulator; (b) five-port circulator; (c) isolator.

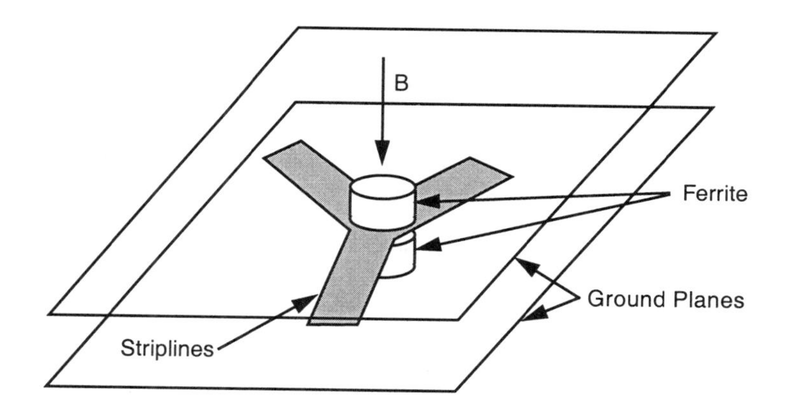

Figure 1.19 A three-port stripline junction circulator. The ferrite disks are biased by a dc magnetic field, provided by a permanent magnet.

out of ferrite, a ceramic-like material having high bulk resistivity but also high permeability and permittivity. Ferrites have the unusual property of becoming nonreciprocal when biased by a dc magnetic field.

A three-port circulator, the most common type, is shown in Figure 1.19. The ports are 120 degrees apart, and the ferrite disks are, in fact, a heavily loaded cylindrical resonator. When a port is excited, two modes are excited, and because of the biased ferrite's nonreciprocity, they propagate around the disks in opposite directions. With careful selection of the ferrite's properties and dimensions, a null can be created at one of the ports, so no power emerges from that port and, as long as the other port is matched, all power emerges from it. (Note that, if the port is not matched, its reflection will emerge from the purportedly isolated port. Clearly, a good port VSWR is essential for good isolation.)

REFERENCES

[1] Wadell, B. C., *Transmission-Line Design Handbook*, Norwood, MA: Artech House, 1991.

[2] Bryant, T., and J. Weiss, "Parameters of Microstrip Transmission Lines and of Coupled Pairs of Microstrip Lines," *IEEE Trans. Microwave Theory Tech.*, Vol. MTT-16, Dec. 1968, p. 1021.

[3] Bahl, I. J., and D. K. Trivedi, "A Designer's Guide to Microstrip Line," *Microwaves*, May 1977, p.174.

[4] March, S., "Microstrip Packaging: Watch the Last Step," *Microwaves*, Dec. 1981, p. 83.

[5] Maas, S., and A. Nichols, *C/NL2 for Windows*, Norwood, MA: Artech House, 1995.

[6] Kirschning, M., and R. Jansen, "Accurate Model of Effective Dielectric Constant of Microstrip With Validity up to Millimeter-Wave Frequencies," *Electron. Ltrs.*, Vol. 18, 1982, p. 272.

[7] Medina, J., A. Serrano, and F. J. Mendieta, "Microstrip Effective Dielectric Constant Measurement and Test of CAD Models up to 20 GHz," *Microwave J.*, March 1993, p. 82.

[8] Atwater, H. A., "Tests of Microstrip Dispersion Formulas," *IEEE Trans. Microwave Theory Tech.*, Vol. MTT-36, March 1988, p. 619.

[9] Kirschning, M., and R. H. Jansen, "Accurate Wide-Range Design Equations for the Frequency-Dependent Characteristics of Parallel-Coupled Microstrip Lines," *IEEE Trans. Microwave Theory Tech.*, Vol. MTT-32, Jan. 1984, p. 83.

[10] Wells, G., and P. Pramanick, "An Accurate Dispersion Expression for Shielded Microstrip Lines," *Int. J. Microwave and Millimeter-Wave Computer-Aided Engineering*, Vol. 5, July 1995, p. 287.

[11] Djordjevic, A., et al., *Linpar for Windows*, Norwood, MA: Artech House, 1996.

[12] Heinrich, W., "Quasi-TEM Description of MMIC Coplanar Lines Including Conductor-Loss Effects," *IEEE Trans. Microwave Theory Tech.*, Vol. MTT-41, Jan. 1993, p. 45.

[13] Howe, H., *Stripline Circuit Design*, Norwood, MA: Artech House, 1984.

[14] Matthaei, G., L. Young, and E. Jones, *Microwave Filters, Impedance-Matching Networks, and Coupling Structures*, Norwood MA: Artech House, 1980.

[15] Yamashita, E., and K. Atsuki, "Strip Line with Rectangular Outer Conductor and Three Dielectric Layers," *IEEE Trans. Microwave Theory Tech.*, Vol. MTT-18, May 1970, p. 238.

[16] Smith, J. I., "The Even- and Odd-Mode Capacitance Parameters for Coupled Lines in Suspended Substrate," *IEEE Trans. Microwave Theory Tech.*, Vol. MTT-19, May 1971, p. 424.

[17] Atwater, H. A., "The Design of the Radial-Line Stub: A Useful Microstrip Circuit Element," *Microwave J.*, Nov. 1985, p. 149.

[18] March, S. L., "Analyzing Lossy Radial-Line Stubs," *IEEE Trans. Microwave Theory Tech.*, Vol. MTT-33, March, 1985, p. 269.

[19] Giannini, F., R. Sorrentino, and J. Vrba, "Planar Circuit Analysis of Microstrip Radial Stub," *IEEE MTT-S Int. Microwave Symp. Digest,* 1984, p. 124.

[20] Gonzalez, G., *Microwave Transistor Amplifiers*, Englewood Cliffs, NJ: Prentice -Hall, 1984.

[21] Collin, R., *Foundations for Microwave Engineering*, 2nd ed., New York: McGraw-Hill, 1992.

[22] Lange, J., "Interdigitated Stripline Quadrature Hybrid," *IEEE Trans. Microwave Theory Tech.*, Vol. MTT-17, 1969, p. 1363.

Chapter 2
Solid-State Devices

Virtually all RF and microwave electronic circuits use one or more of three general types of devices: Schottky-barrier diodes, junction transistors, or field-effect transistors (FETs). Within these broad categories are many different types of devices: a wide variety of Schottky-barrier diodes, optimized for either low cost or high performance; bipolar-junction transistors (BJTs); heterojunction bipolar transistors (HBTs); and various types of FETs, including metal-epitaxial semiconductor FETs (MESFETs), high-electron-mobility transistors (HEMTs), metal-oxide semiconductor FETs (MOSFETs), and junction FETs (JFETs).

These devices have distinctly different characteristics, so in most cases the appropriate device for a particular circuit is obvious. The choice of a device also may be colored by the available technologies and, above all, cost.

Table 2.1 lists the solid-state devices described in this chapter. The suggestions for applications and frequency ranges are weak; often there are good reasons to use a device outside its optimum frequency range or for applications where it might not, at first inspection, seem appropriate. This information is valid for late 1997; it may be obsolete by the time you read this.

2.1 SCHOTTKY-BARRIER DIODES

The Schottky-barrier diode is the cockroach of microwave technology: it is has been around since the beginning and is impossible to exterminate. Schottky-barrier diodes existed before any other microwave electronic devices and will be around long after all the others are gone. The galena detector used in crystal radios in the 1920s is a type of crude Schottky diode, and the earliest microwave mixers used point-contact diodes, a type of Schottky-barrier diode only slightly less crude than a chunk of galena and an adjustable wire contact. Even now, with the existence of microwave transistors, Schottky-barrier-diode mixers and frequency multipliers still have several advantages over their active counterparts and still are widely used.

Table 2.1 Solid-State Devices Described in This Chapter

Device	Frequency Range[a]	Uses and Characteristics
Schottky-barrier diode	The RF to the submillimeter range.	Mixers, modulators, and detectors; occasionally used for frequency multipliers and switches.
Varactor and step-recovery diode (SRD)	*Pn*-junction varactor: to ~50 GHz; Schottky varactor: to several hundred GHz; SRD: to ~18 GHz.	Frequency multipliers; SRDs are used to generate fast pulses and for high-order frequency multiplication.
Bipolar junction transistor (BJT)	Usually X band and below; millimeter-wave BJTs have been made.	Small-signal amplifiers; not low noise. Fast digital circuits. Good power devices. Low 1/f noise makes them ideal for low-noise oscillators.
Heterojunction bipolar transistor (HBT)	Some types may have gain at 60 GHz; practical limits are around 30 GHz.	Power amplifiers, low-noise oscillators. Fast analog circuits. MESFETs and HEMTs have much better noise figures, but HBTs have lower 1/f noise, making them preferable for oscillators.
Junction field-effect transistor (JFET)	Up to the VHF/UHF range.	Low-cost, moderately low-noise applications in amplifiers, mixers, oscillators, and switches.
Metal-epitaxial semiconductor FET (MESFET)	Microwave workhorse up to 26 GHz. Can go higher, but HEMTs are usually preferred.	Amplifiers, oscillators, mixers, modulators, frequency multipliers, control components; in short, everything. Bipolar devices have lower 1/f noise and are preferred for oscillators.
High-electron-mobility transistor (HEMT)	Highest frequency device available; over 200 GHz.	Much the same as MESFETS; best suited for small-signal, low-noise uses, but power devices are possible.
Metal-oxide semiconductor FET (MOSFET)	Up to ~6 GHz for advanced technologies; 2-3 GHz more common.	Analog, digital, and RF Si IC applications. MESFETs and HEMTs have much lower noise figures at microwave frequencies.

a. As of late 1997. These ranges change yearly.

Schottky-barrier diodes can be used in frequency multipliers, as well as mixers, and in many types of detectors and wave-shaping circuits. The uses of these diodes in such applications are covered in later chapters.

2.1.1 Fundamental Properties

A Schottky barrier is a metal-to-semiconductor junction that can rectify. A Schottky-barrier diode is simply a Schottky junction used as a diode. These devices are used most often in mixers and detectors, but they also are used in resistive-diode frequency multipliers, millimeter-wave reactive frequency multipliers, and in other types of circuits where fast-switching diodes are needed.

Schottky-barrier diodes, or simply Schottky diodes, are about as simple in structure as microwave electronic devices get. The structural simplicity is deceiving, however, because a great amount of effort has been applied to the perfection of these junctions. In fact, for many years, the development of Schottky diodes for millimeter-wave mixers was almost exclusively the driving force behind improvements in microwave device technology.

Why does a metal-to-semiconductor junction rectify? A better question is why *don't* some junctions rectify? Schottky junctions rectify because the metal's work function is greater than the semiconductor's. This creates an energy barrier between the semiconductor and the metal, which decreases when the junction is forward biased and increases when the junction is reverse biased. To achieve a contact that doesn't rectify, called an *ohmic contact*, we need a metal whose work function is smaller than the semiconductor's. Most practical metals and semiconductors don't have this property. As a result, making a good ohmic contact takes a lot more effort than simply selecting materials.

Why are Schottky diodes used instead of *pn*-junction diodes? In a Schottky diode, current consists entirely of majority carriers, invariably electrons. For this reason, a Schottky diode is called a *majority-carrier device*. In a *pn*-junction diode, conduction is dominated by minority carriers. When the *pn* diode is forward biased, a substantial amount of minority charge is stored in the junction, and, if the diode is suddenly reverse biased, the charge must be removed before the diode can turn off. This process is relatively slow and prevents the use of such diodes as rectifiers at high frequencies. *pn*-junction devices, however, can be used as voltage-variable capacitors, called *varactors*, and may have advantages over Schottky diodes in such applications. In these devices, charge storage is actually a benefit.

Many different types of Schottky diodes are used in RF and microwave circuits. They can be realized in virtually any type of semiconductor, although only silicon and GaAs are commonly used for discrete devices. The great majority of commercially available diodes are silicon, as is the widest variety of types of devices and packages. Nevertheless, we can generalize a bit about such devices. All Schottky diodes have the general structure shown in Figure 2.1. The diode is built on a high-conductivity *n*-type substrate or at least has a high-conductivity layer underneath it.

n material is used exclusively for high-frequency devices; *p* material is distinctly inferior to *n* in both GaAs and silicon. Above the substrate is a relatively thick (a few microns) *n*+ buffer layer, whose purpose is to separate the epitaxial layer, or *epilayer*, from the impurities and imperfections in the substrate. The epitaxial layer is grown on top of the buffer. This is quite thin, 1,000 to 2,000 Å, and is much more lightly doped. Finally, a metal layer, the *anode*, is deposited on the epilayer. The metal, which establishes the area of the junction, usually is in the form of a circular dot. A variety of metals can be used; the selection of metal, anode shape and area, doping density, and epilayer thickness are about the only degrees of freedom available for adjusting the diode's characteristics.

An ohmic contact, the *cathode*, is formed on the *n*+ substrate. It can be formed on the underside of the diode's substrate or, by removing the epilayer and exposing the buffer, on the top. If the ohmic contact is made at the upper surface of the diode, the substrate need not be high conductivity, although an *n*+ region under the epilayer is still needed. (This is how diodes in monolithic circuits are made.) The ohmic contact usually is an alloy—gold-germanium is common—and is gold plated.

Figure 2.1 shows a few other details. The top of the substrate has an oxide layer. This passivates the surface, preventing contamination from foreign substances in the diode's environment, and defines the anode during processing. Finally, the anode is gold plated to facilitate making the anode's electrical connection.

2.1.2 Electrical Characteristics

The contact between the anode metal and the semiconductor causes electrons from the semiconductor to move to the surface of the metal. This depletes the

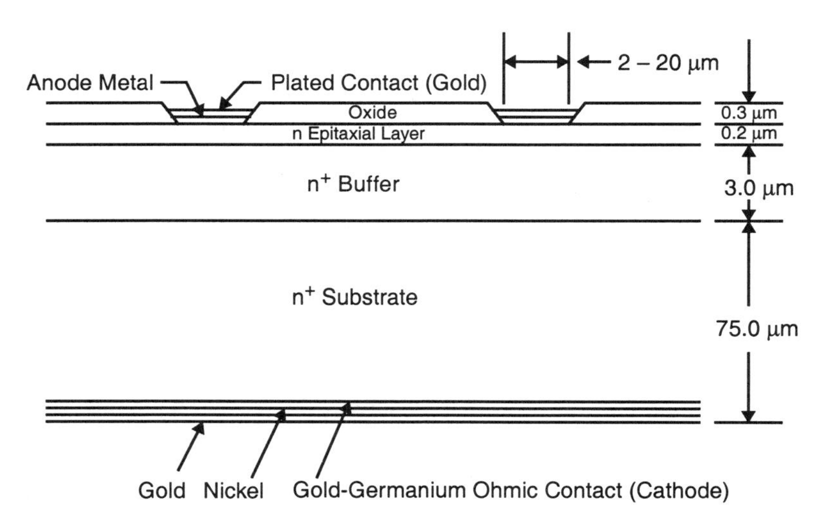

Figure 2.1 Cross section of a Schottky-barrier diode.

semiconductor of electrons in a region underneath the anode, predictably called a *depletion layer*. The movement of charge creates a negative charge on the surface of the anode and a positive charge (from ionized dopant atoms) in the semiconductor. The electric field created by this charge dipole opposes further movement of charge, creating an equilibrium in the metal-semiconductor system. Applying a voltage to the junction increases or decreases this field, and charge moves between the metal and the semiconductor to reestablish equilibrium, widening or narrowing the depletion region in the process. In this way—as charge moves between the two "poles" (the anode surface and the semiconductor)—the junction behaves as a nonlinear capacitor.

When the device is in equilibrium, and no external voltage is applied to the junction, electrons move as easily from the metal to the semiconductor as from the semiconductor to the metal, resulting in no net current. However, if a positive voltage is applied to the anode, the barrier is lowered, allowing more electrons to cross it from the semiconductor to the metal, and a current results. This current is a strongly nonlinear function of the voltage.

From this description it is apparent that the Schottky junction behaves as a nonlinear resistor in parallel with a nonlinear capacitor. A Schottky junction is a device—one of the few microwave devices—that can be described adequately for almost all purposes by simple, closed-form equations. The current-voltage (I/V) characteristic of the junction is

$$I_j(V_j) = I_s \exp\left(\frac{qV_j}{\eta KT}\right) \tag{2.1}$$

where I_j is the junction current and V_j is the junction voltage.

These parameters require some explanation. I_s is often called the *reverse saturation current*, since the equation implies that $I_j(V_j \rightarrow -\infty) = I_s$. In fact, this is a very small constant, 10^{-20} to 10^{-8} A, depending on the type of diode, and the reverse current is rarely so low. I_s adjusts the forward I/V characteristic: I_s affects the "knee" of the characteristic, the voltage at which the junction current has some standard value, usually 10 μA or 100 μA. η is the ideality factor, a parameter that accounts for the nonideality of the junction. In a good diode, η is between 1.05 and 1.25; an ideal junction has $\eta = 1.0$. The other terms are familiar physical constants: q is electron charge (1.6×10^{-19} C.), K is Boltzmann's constant (1.37×10^{-23} J/K) and T is absolute temperature in Kelvins.

The capacitance-voltage (C/V) characteristic is

$$C_j(V_j) = \frac{C_{j0}}{\sqrt{1 - \frac{V_j}{\phi}}} \tag{2.2}$$

where C_{j0} is the junction capacitance at zero voltage and ϕ is the built-in voltage of the junction. ϕ is the potential difference between the semiconductor and the metal anode when no external voltage is applied; it is the quantity obtained by integrating the electric field across the depletion region. ϕ depends on the type of metal and semiconductor used in the junction. For silicon diodes ϕ typically is 0.6V; for GaAs diodes $\phi \cong 0.75$V. This expression is valid as long as the epilayer is thick enough to prevent the depletion region from "punching through" to the substrate at high reverse voltages. When the depletion region does punch through, the capacitance variation with voltage suddenly becomes very weak. This phenomenon is used in some types of frequency-multiplier diodes, to minimize the multiplier's output-power variation, and in *Mott diodes*, which have very low noise in cooled millimeter-wave mixers.

Equation (2.2) has an obvious difficulty: $C_j(V_j) \rightarrow \infty$ as $V_j \rightarrow \phi$. This is more of a paradox than a real problem. At $V_j = \phi$ the depletion region disappears, so the depletion charge Q_d is zero. The capacitance, defined as dQ_d / dV_j is indeed infinite, but this definition of capacitance is valid only for infinitesimal RF junction voltage. Properly designed nonlinear-circuit simulators circumvent this problem by using increments of charge, instead of the capacitance function, to estimate current.

Figure 2.2 shows the equivalent circuit of the intrinsic diode; that is, the junction alone, not including package or other parasitics. The circuit includes a constant series resistance, R_s, as well as the $C_j(V_j)$ and $I_j(V_j)$ elements. The series resistance, which is an unavoidable component of any diode, comes from the undepleted epilayer under the junction and may have a small component from the ohmic contact and substrate resistances as well. Although this resistance is weakly nonlinear, we normally treat it as a linear element.

2.1.3 Practical Schottky Diodes

Many types of Schottky diodes are available and can be obtained in a wide variety of packages. The best selection is in silicon; GaAs devices, being considerably more expensive than silicon, are normally reserved for high-performance applications and

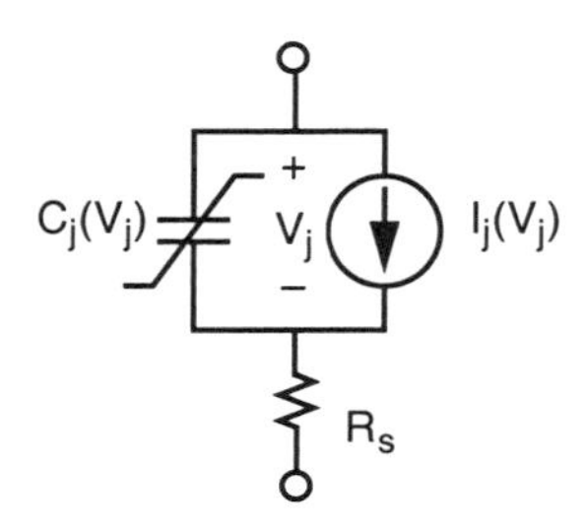

Figure 2.2 Equivalent circuit of the junction of a Schottky-barrier diode. This circuit describes only the intrinsic junction; additional elements may be needed to describe the parasitic elements of some types of diodes. See (2.1) and (2.2) for the I / V and C / V expressions.

are available in a narrower variety of packages. It makes no sense to put a high-performance diode in a low-performance package.

By far, the most widely used diodes are plastic-packaged devices and beam-lead devices. The most common type of "plastic" package is really a tiny disk of ceramic with metal leads. The chip is mounted on the ceramic, wire bonds connect it to the package's ribbon leads, and a blob of epoxy covers the whole thing. Such a package is not hermetic and cannot be used for high-reliability applications. Hermetic ceramic-metal packages, which are considerably more expensive, can be used for high-reliability components.

A beam-lead diode is a chip that has integral gold ribbons, which are formed in the diode-fabrication process. The main disadvantages of beam-lead diodes are their top-side ohmic contact, which increases the series resistance, and a troublesome overlay parasitic capacitance between the anode ribbon and the substrate, which is in parallel with the junction. This overlay capacitance can be reduced in conventional beam-leads only at the expense of making the device very fragile. Beam-lead diodes have a well-deserved reputation for fragility. Fortunately, new types of beam-lead diodes offer greatly reduced overlay parasitics and greater ruggedness at little additional expense.

A type of device deservedly gaining in popularity is the so-called leadless beam-lead device. This device has thick, integral mounting pads and can be mounted upside-down on a substrate. Unlike beam-leads, these devices are rugged enough for machine insertion into circuits. They are attached by solder or conductive adhesives. Finally, diodes are available in a variety of small, epoxy surface-mount packages. These have relatively large parasitics and therefore are not suitable for high frequencies.

Diodes are available as single devices, "tees" (two diodes) or "quads" (four diodes) in a single package or on a single chip. (The need for such devices will be clarified in Chapters 3 and 4.) Silicon diodes are also available in various barrier heights; a low-barrier diode has a knee in its I/V characteristic around 0.3V; a high-barrier device around 0.6V. The low-barrier devices generally operate at lower local oscillator (LO) power in mixers or at lower input levels in resistive frequency multipliers, but they are not as good at large-signal handling.

Creating an equivalent circuit of a package is not a simple matter. Again, the usual technique is to measure the package's S parameters, or to calculate them by an electromagnetic simulator, and to fit them numerically to the equivalent circuit. In some cases elements in the package model can be isolated by measuring the package with internal nodes short-circuited or open-circuited. Occasionally the diodes' manufacturer can provide package models. Some manufacturers do a better job of this than others; this is one place where skepticism is in order.

2.1.4 Diode Selection

Equations (2.1) and (2.2) show that the circuit designer does not have many degrees of freedom in selecting a diode, and a moment's consideration indicates that many of the parameters are linked. The only parameters available for adjusting a diode's characteristics are I_s, C_{j0}, and R_s; all others are physical constants, are normally minimized (η), or are so strongly linked to the materials or device-fabrication process that they are not really adjustable (ϕ).

I_s and C_{j0} are roughly proportional to the anode's area, and R_s is inversely proportional. Thus, the quantity $R_s C_{j0}$ is roughly constant with anode area, and we can define a figure of merit, f_c, called the diode's *cutoff frequency*,

$$f_c = \frac{1}{2\pi R_s C_{j0}} \tag{2.3}$$

Cutoff frequencies can be startlingly high: 2,000 GHz is quite common, and 4,000 GHz for high-performance diodes is not unheard of. Remember, this is just a figure of merit; a 2,000-GHz diode cannot necessarily be used successfully in a 2,000-GHz mixer!

These parameters also are linked through the physical characteristics of the device. For example, as doping density is increased, R_s decreases, C_{j0} increases, and reverse-breakdown voltage decreases. GaAs devices have higher electron mobility than silicon, so they can be doped more lightly, achieving higher cutoff frequencies and higher breakdown voltages. I_s depends strongly on doping density and the metal-semiconductor combination; these are selected to provide high, medium, or low barrier heights.

The diode manufacturer selects the anode and epilayer parameters to provide a certain barrier height and to optimize f_c. Diodes having a number of anode areas are then produced, resulting in diodes with low C_{j0} but relatively high R_s, or higher C_{j0} and lower R_s. A diode then is selected by the circuit designer to have the best R_s-C_{j0} trade-off. This trade depends on the type of circuit; we examine it in more detail in Chapters 3 and 4 when we discuss specific circuits.

One of the worst ways to select a diode is on the basis of its performance characteristics, usually noise figure and conversion loss, listed in a diode manufacturer's catalog. These specifications are meaningless. Manufacturers measure these quantities in standard test fixtures. Unfortunately, they are really measuring the test fixture, not the diode! Diode mixers and frequency multipliers are *circuit limited*; the circuit, not the diode, generally limits the performance. Theoretically, any diode is capable of far better performance than any of us will ever see. We never achieve this theoretical capability because the circuit needed to achieve optimum performance is not realizable in any practical manner.

2.1.5 Diodes for Monolithic Circuits

In principle, Schottky diodes can be fabricated in any monolithic technology. The variety of diodes available in such technologies is dictated less by intrinsic limitations of the technology than by the need for process compatibility with FETs, bipolar transistors, or other solid-state devices.

Various monolithic technologies have their own compatible diode structures. Bipolar technologies may use a metal deposition on a collector junction, FETs may use a gate-to-channel junction, and some silicon technologies do not include Schottky diodes at all. Occasionally a foundry develops an independent diode process, one that is not a child of a FET or HBT technology.

Figure 2.3 shows a mesa diode. Such diodes are used by the few of us lucky enough to have an uncompromised diode technology. The semiconductor portion of the diode consists of an *n*+ layer capped by an *n* epilayer. The epilayer is removed on three sides of the anode, allowing access to the *n*+ layer for the ohmic cathode contact. To minimize overlay capacitance, the anode connection is formed by an air bridge; to minimize series resistance, the anode often is relatively long and narrow. However, it must not be so narrow that the anode metal itself has appreciable resistance.

Many monolithic technologies use the gate-to-channel junction of a MESFET (Section 2.7) as a Schottky diode. The high gate resistance of such diodes causes the series resistance to be high. Furthermore, the voltage drop along the gate causes the current density to be greater near the gate-connection pad than at the end farthest from the pad. The result is a poor ideality factor and a relatively high junction capacitance.

2.1.6 Diode Measurements

Measuring a diode's I / V characteristic is extraordinarily easy, in part because any decent Schottky diode exhibits an accurate exponential characteristic over several decades of current. To find I_s and η we first determine the slope of the I / V curve in millivolts per decade of current. From (2.1) we find that the change in voltage, ΔV, giving a decade change in current is

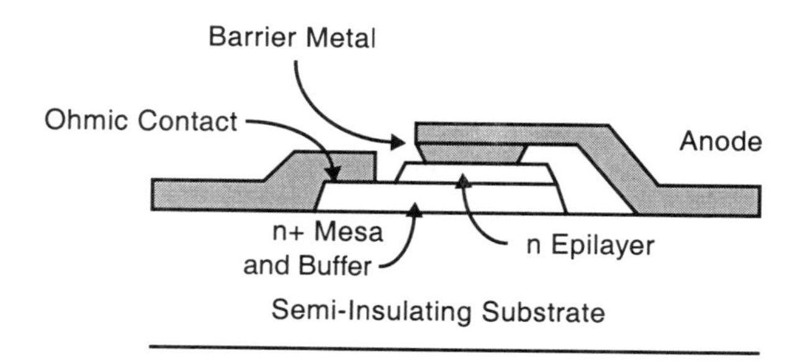

Figure 2.3 Mesa diode for monolithic circuits. The cathode ohmic contact surrounds the anode on three sides and minimizes the series resistance. The air bridge minimizes overlay capacitance.

$$\Delta V = \frac{\eta KT}{q\log(e)} \tag{2.4}$$

where e is the base of natural logarithms. This comes to 57.8 η mV/decade at room temperature (295K). Knowing ΔV, we solve for η. Once η is known, I_s can be found easily from (2.1).

Finding R_s is a little trickier. R_s is on the order of 10Ω. When the junction current exceeds 1 mA, a few millivolts are dropped across R_s, and the slope of the I / V curve deviates from the ideal. R_s is

$$R_s = \frac{\delta V}{I_j} \tag{2.5}$$

where δV is the voltage deviation from the ideal curve at I_j, the diode current. Figure 2.4 summarizes these measurements.

Because it is such a small capacitance, C_{j0} is considerably more difficult to measure. Several methods used in the past are outlined in Reference [1]. More recently, since accurate automatic network analyzers have become available, C_{j0} is now measured in the same way as the capacitances of transistor models: the device is placed in a test fixture, its S parameters or reflection coefficient are measured, and

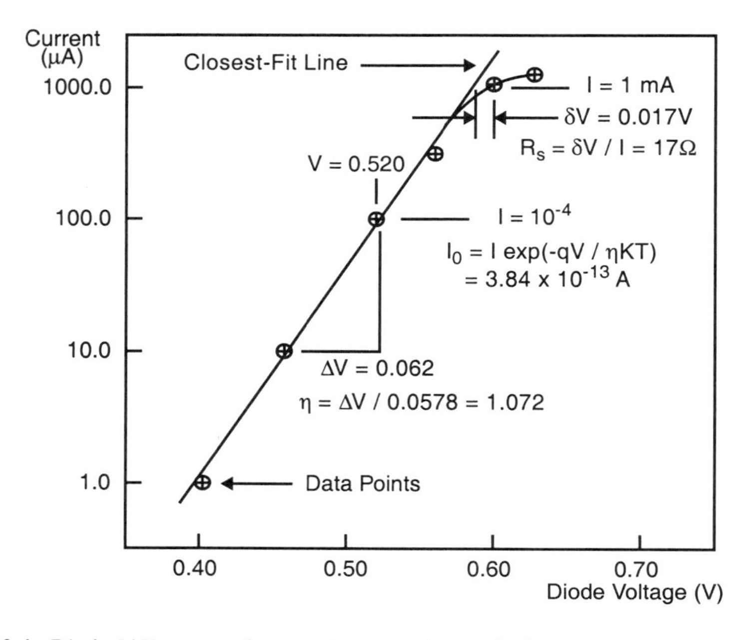

Figure 2.4 Diode I / V curve and parameter-extraction methods.

these are fit to the equivalent circuit by numerical optimization. When carefully performed, this process can account for packaging and overlay parasitics as well as C_{j0}.

2.2 VARACTOR AND STEP-RECOVERY DIODES

Varactors[1] are diodes used primarily for their nonlinear junction capacitance. Most are *pn*-junction devices, but Schottky-barrier diodes are sometimes used as well. Varactors are used in frequency multipliers and as tuning elements in filters and oscillators. Schottky varactors are used only for millimeter-wave applications, where the capacitances of diffused *pn*-junction devices are too high.

Step-recovery diodes (SRDs) are *pn*-junction diodes. Unlike conventional varactors, they use charge storage in forward conduction to produce a nonlinear reactance. This nonlinearity is much stronger than the depletion capacitance and can be used to generate fast pulses or high-order frequency multiplication.

2.2.1 Fundamental Properties

Interestingly, charge storage—the property that makes a *pn*-junction diode useless for mixers—makes it very useful as a varactor. When a varactor is used as a frequency multiplier, the strength of its nonlinearity determines its efficiency. Unfortunately, the depletion-capacitance nonlinearity is fairly weak, and this makes it difficult to design a really good reactive frequency multiplier. One way to increase the nonlinearity is to drive the diode hard enough to inject charge into its junction. Overdriving the varactor in this manner does not result in forward conduction. For conduction to occur, the charge carriers have to recombine; because of the long lifetime of the injected minority carriers, most do not recombine, and very little conduction occurs. While a conventional *pn*-junction varactor makes only modest use of the charge-storage capacitance, an SRD uses it almost exclusively; this is the main difference between these two devices. Schottky varactors do not store charge at all. If driven into forward conduction, the Schottky diode absorbs energy and the multiplier's efficiency decreases.

2.2.2 Electrical Characteristics

The Schottky C/V characteristic was given in Equation (2.2). That equation applies only to a uniformly doped diode in which the epitaxial layer is never fully depleted. For better or worse, however, most diffused *pn*-junction varactors are not uniformly doped; the doping, especially in the vicinity of the junction, usually is nonuniform. Still, for purposes of design, the diode is assumed to be uniformly doped and (2.2) applies. This assumption does not make much difference, in the final analysis, in the

1. The term is an elision of the words *variable* and *reactor*; a varactor is a voltage-variable reactance.

design of a frequency multiplier.

Some varactors, called *punch-through* or *dual-mode varactors*, are designed to be fully depleted of charge at a modest reverse voltage. Because the capacitance ceases to vary when the junction is fully depleted, these varactors have a slightly weaker nonlinearity when strongly driven, and the output power varies less with input power. This costs some efficiency, but the flat P_{out} / P_{in} characteristic is valuable in many applications.

Hyperabrupt varactors have a much stronger C / V nonlinearity. They are used for tuning filters and oscillators; they are not used in frequency multipliers. The series resistance of a hyperabrupt varactor is higher than that of a conventional varactor, and in frequency multipliers this would reduce efficiency more than the stronger nonlinearity would increase it.

2.2.3 Equivalent Circuit

The equivalent circuit of a conventional varactor is the same as that of the Schottky diode (see Figure 2.2). As with the Schottky, this circuit describes the junction alone. It may be necessary to add package or other parasitics to it.

Although Figure 2.2 applies to SRDs as well, the SRD equivalent circuit usually is simplified. Because it operates only in strong forward bias or at a large reverse bias, the SRD can be treated as a short circuit when forward biased and a small capacitance when reverse biased. (This moderately brainless model is used in the description of SRD multipliers in Chapter 4.)

2.3 BJTS

BJTs are realized only in silicon. Si BJTs are used in a wide a range of applications, in discrete and monolithic forms, from dc to the lower microwave range. To differentiate them from heterojunction bipolar transistors, BJTs sometimes are called *homojunction bipolar transistors*. Electronic technology runs on acronyms, however, and since both heterojunction and homojunction bipolar transistors share the same initials, we simply call the latter devices bipolar junction transistors, or *BJTs*.

BJTs are not practical in GaAs or InP because diffusion of *p* dopants in GaAs is difficult to control. *p* material also has low mobility, and the lightly doped bases necessary for good current gain would have very high base resistance. The combination of high base resistance and poorly controlled base width makes a conventionally fabricated GaAs BJT poorer in performance than a silicon one.

GaAs BJT bases probably could be fabricated by molecular-beam epitaxy. Once we decide to use such costly techniques, however, it makes sense to go all the way to a heterojunction device. GaAs and InP HBTs occupy the technological high ground; SiGe HBTs are used as high-performance, low-cost devices. Si BJTs are reserved largely for more the prosaic applications, below a few GHz, requiring more modest performance and low cost. HBTs are discussed further in Section 2.4.

2.3.1 Fundamental Properties

In many ways a BJT operates like a *pn*-junction diode with an extra terminal. The current in the *pn* junction is controlled by the voltage between its terminals (the *base* and the *emitter*) but, unlike a *pn*-junction diode, the resulting current is collected by a third terminal, called (believe it or not) the *collector*. The voltage between the base and emitter terminals controls the current in the collector, resulting in a very useful phenomenon called *amplification*.

Figure 2.5 shows the structure of a BJT. The emitter, base, and collector regions are long and narrow, to provide adequate emitter area for high peak current, while minimizing the resistance of the base region. Most practical BJTs, especially power devices, consist of many such cells connected in parallel.

Figure 2.5 shows an *NPN* device. In microwave BJTs the emitter and collector invariably are *n* material, and the base is a very thin region of *p* material. The *p* base is lightly doped, and the *n* emitter is heavily doped. This structure causes the junction current to consist almost entirely of electrons injected into the base; it also minimizes hole injection into the emitter, which contributes to base current but not collector current. The result is high current gain, usually indicated by the Greek letter β:

$$\beta = \frac{I_C}{I_B} \tag{2.6}$$

Because the base is very thin, electrons from the emitter cross the base quickly and have little opportunity to recombine with holes in the base. Instead, they are swept into the collector, which is biased to a relatively large positive voltage, at least a few volts. Thus, the voltage applied to the base-emitter junction controls a current in the collector. Like the *pn*-junction diode it mimics, the BJT has an exponential I/V characteristic.

There are a few complications in this rosy picture of BJT operation. First, the base current is small, but not zero, caused not only by hole injection into the emitter but also by a small amount of recombination in the base. These phenomena limit the current gain. Second, because of the finite transit time of the electrons, charge is stored in the base. This is the charge of the relatively large base-to-emitter capacitance, which, in combination with the high resistance of the lightly doped base, limits the BJT's performance significantly. Third, the BJT can operate in reverse; if the base-collector junction becomes forward biased, it can act, in a fashion, like the base-emitter junction. The light collector doping, however, causes the current gain to be very low. Finally, like other devices, BJTs have various parasitic elements, including depletion capacitances at both junctions and series resistances at all three terminals. These have a strong effect on the performance of the device.

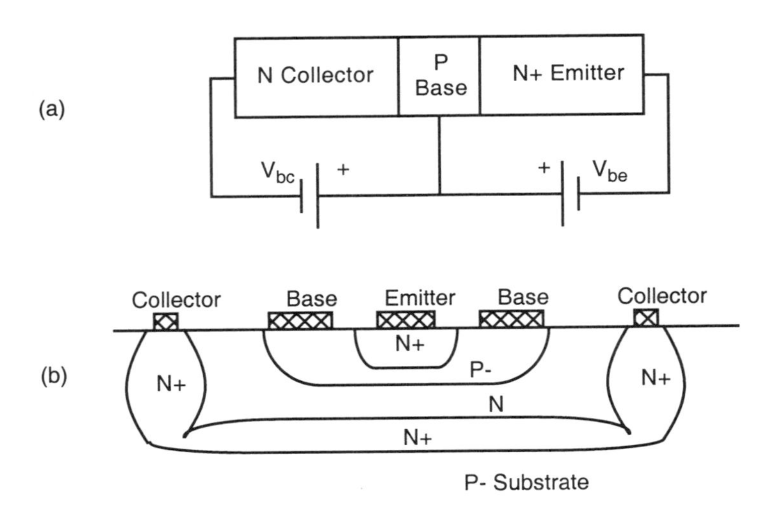

Figure 2.5 (a) Idealized NPN BJT. Virtually all RF and microwave devices have an NPN structure. In normal operation V_{bc} is negative; that is, the collector-to-emitter voltage is positive. (b) A planar diffused BJT structure. To minimize collector-to-substrate capacitance, these devices have oxide (SiO2) isolation alongside the collector *n*+ "sinkers," and the *p*-layer is biased to the largest negative voltage available in the circuit.

2.3.2 Electrical Characteristics

Static I/V Characteristic

The forward *I/V* characteristic of the BJT is established by the *I/V* characteristic of the base-to-emitter *pn* junction:

$$I_{cf} = I_{sf}\left[\exp\left(\frac{qV_{be}}{KT}\right) - 1\right] \tag{2.7}$$

where V_{be} is the base-to-emitter voltage and I_{cf} is the forward collector current. We must, of course, include the possibility of reverse operation:

$$I_{cr} = I_{sr}\left[\exp\left(\frac{qV_{bc}}{KT}\right) - 1\right] \tag{2.8}$$

where V_{bc} is the base-to-collector voltage and I_{cr} is the reverse collector current. The principle of reciprocity demands that the current parameters I_{sf} and I_{sr} be equal. The total collector current then becomes

$$I_c = I_{cf} - I_{cr} = I_{sf}\left[\exp\left(\frac{qV_{be}}{KT}\right) - \exp\left(\frac{qV_{bc}}{KT}\right)\right] \tag{2.9}$$

The collector-to-emitter voltage V_{ce} is

$$V_{ce} = V_{be} - V_{bc} \tag{2.10}$$

and substituting (2.10) into (2.9) gives

$$I_c = I_{sf}\exp\left(\frac{qV_{be}}{KT}\right)\left[1 - \exp\left(\frac{-qV_{ce}}{KT}\right)\right] \tag{2.11}$$

Equation (2.11) describes the collector I/V characteristic of the BJT. The first of its two terms establishes the peak collector current, which occurs when V_{ce} is large. The second sets the shape of the collector-current curve as a function of V_{ce}.

Collector I/V curves usually are plotted with the base current, not V_{be}, as a parameter. An example of a set of collector I/V curves is shown in Figure 2.6.

The transconductance G_m of the BJT is an important quantity for small-signal applications. It is simply

$$G_m = \left.\frac{dI_c}{dV_{be}}\right|_{V_{ce} \gg \frac{KT}{q}} \cong \frac{q}{KT}I_c \tag{2.12}$$

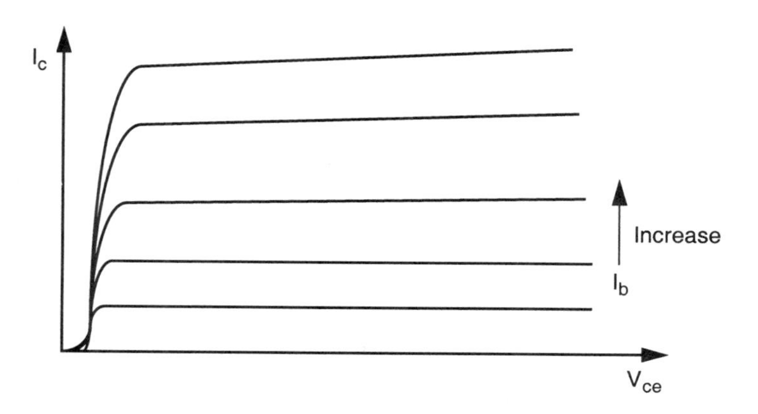

Figure 2.6 Collector I/V curves of a small-signal BJT.

Capacitances

In a BJT both junctions have capacitances. Each junction capacitance has two components: a depletion component and a diffusion component. The depletion component is given by the well-known diode-capacitance equation,

$$C_{je} = \frac{C_{je0}}{\sqrt{1 - \frac{V_{be}}{\phi_{be}}}} \tag{2.13}$$

$$C_{jc} = \frac{C_{jc0}}{\sqrt{1 - \frac{V_{bc}}{\phi_{bc}}}} \tag{2.14}$$

The diffusion capacitance is caused by charge storage in the base. If the base transit time in normal, forward conduction is τ, the charge stored in the base Q_b is simply τI_c. The capacitance is

$$C_{be,\tau} = \frac{dQ_b}{dV_{be}} = \frac{I_{sf}q}{KT}\tau \exp\left(\frac{qV_{be}}{KT}\right) \tag{2.15}$$

The total base-to-emitter capacitance is the sum of these quantities,

$$C_{be} = C_{je} + C_{be,\tau} \tag{2.16}$$

In reverse conduction (which normally does not occur in RF or microwave devices) there is a diffusion capacitance associated with the base-to-collector junction. The form is analogous to (2.15), and C_{bc}, the total base-to-collector capacitance, is the sum of the depletion and diffusion components.

At normal collector currents, the diffusion component is the dominant part of the base-to-emitter capacitance. Clearly, minimizing the width of the base minimizes τ, and therefore minimizes $C_{be,\tau}$. Minimizing the base width also increases the base resistance, however, so high-frequency performance is not necessarily improved. HBTs neatly circumvent this problem by using a heterostructure, instead of a *pn* junction, for the base-to-emitter junction. This creates a large barrier to prevent hole injection from the base, even when the base is heavily doped and the base resistance is small. More on this later.

2.3.3 BJT Figures of Merit

Two important figures of merit can be defined for the BJT. The less important of the two, for RF and microwave purposes, is the current-gain cutoff frequency, f_t. f_t is the frequency at which the extrapolated magnitude of the short-circuit current gain (which is the same as the hybrid parameter H_{21}) is 1.0. A good approximation for this quantity is

$$f_t = \frac{G_m}{2\pi C_{be}} = \frac{1}{2\pi\tau} \tag{2.17}$$

In (2.17) we have assumed that the diffusion capacitance is dominant. It is important to remember that this quantity does not account for the effects of base resistance; base resistance is a significant parasitic in high-frequency BJTs. Nevertheless, the simple way in which this quantity relates three important parameters makes it very useful.

The second figure of merit is f_{max}, the available-gain bandwidth product, or, in other terms, the extrapolated frequency at which the available gain of the BJT reaches 1.0. It sometimes is called the *maximum frequency of oscillation*. Both f_{max} and f_t are extrapolated from the part of the curve of gain versus frequency that exhibits a slope of 6 dB/octave. These frequencies are often so high that they cannot be measured directly.

2.3.4 Other Parasitics

BJTs have ohmic contact resistances at all terminals. Because the base region is lightly doped, they have significant base resistance as well. BJTs also have a small capacitance associated with the collector-to-substrate junction; because this junction is strongly reverse biased, its capacitance usually is small.

2.3.5 Large-Signal Equivalent Circuit

Figure 2.7 shows the large-signal equivalent circuit of the BJT. This nonlinear equivalent circuit accounts for both forward and reverse operation. In high-frequency circuits, especially small-signal ones, reverse operation normally does not occur; in power circuits, however, the BJT can be pushed into its saturation region, where those elements are necessary.

2.3.6 Small-Signal Equivalent Circuit

A small-signal equivalent circuit can be obtained by linearizing the large-signal circuit of Figure 2.7 in the vicinity of the bias point. The resulting "tee" circuit topology is then converted to the so-called *hybrid-pi* circuit shown in Figure 2.8. The hybrid-pi circuit is generated in two steps. First, the nonlinear elements of the large-signal circuit are linearized in the vicinity of the bias point. The large-signal controlled source becomes a linear transconductance,

$$i_c = G_m v_{be} \tag{2.18}$$

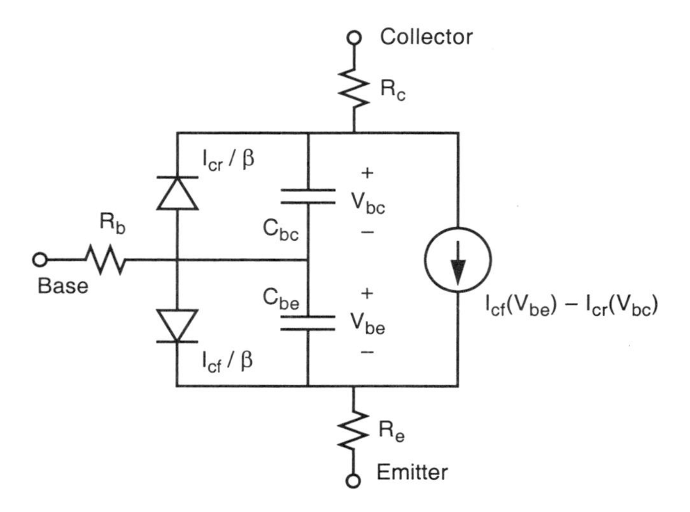

Figure 2.7 Large-signal equivalent circuit of a BJT. C_{be} and C_{bc} have both depletion and diffusion components. Often a second pair of diodes is included to account for base leakage current at low base-to-emitter voltages.

where G_m is given by (2.12), evaluated at the bias point. The base-to-emitter diode becomes a resistor, R_{je}, whose value is found by differentiating (2.7):

$$G_{je} = \frac{1}{R_{je}} = \frac{q}{KT} I_c \tag{2.19}$$

and I_c is, again, the bias current. The resulting tee circuit can be difficult to analyze. Fortunately, we can approximate it by a pi circuit (hence the name *hybrid pi*). To do so we multiply the emitter-loop resistances by the current gain, β, and move them to the base loop; the transconductance must be modified as well. Finally, we have

$$\begin{aligned} R_\pi &= \beta(R_{je} + R_e) \\ G_m &\rightarrow \frac{G_m}{1 + G_m(R_{je} + R_e)} \end{aligned} \tag{2.20}$$

2.3.7 Gummel-Poon Model

The previous description of the BJT is, in fact, the Ebers-Moll [2] model of the device. This model is valid for many ordinary purposes. The Gummel-Poon model [3] accounts for several important phenomena not included in Ebers-Moll. The Gummel-Poon model is sometimes called a *charge-control model*, because it describes the collector current as a function of a quantity, q_b, the normalized

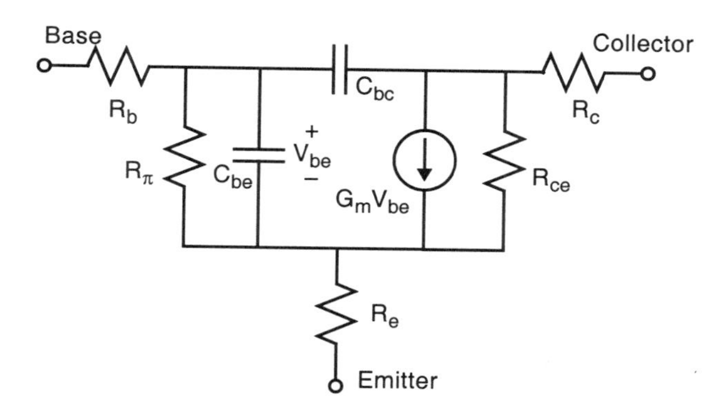

Figure 2.8 Small-signal "hybrid-pi" equivalent circuit of a BJT. This is a linearization of the large-signal model, with the base-to-emitter junction resistance and the emitter parasitic resistance moved to the base circuit. This approximation simplifies analysis of the circuit.

majority charge stored in the base. The user, however, need not be aware of this; the charge-control operation is described by a relatively small number of parameters in the model. If these special parameters are eliminated, the Gummel-Poon model reduces to the Ebers-Moll model.

The SPICE[2] implementation of the Gummel-Poon model includes some further enhancements and has become more of a standard than the model in the original paper. That model is described here. The topology of the model is almost identical to that of Figures 2.7 and 2.8; the circuits are shown in Figures 2.9 and 2.10.

High-Level Injection

A more precise analysis of the BJT gives the following relation for collector current, instead of (2.9):

$$I_c = \frac{C}{p}\left[\exp\left(\frac{qV_{be}}{\eta_f KT}\right) - \exp\left(\frac{qV_{bc}}{\eta_r KT}\right)\right] \tag{2.21}$$

where p is the majority-carrier concentration in the base, C is a constant, and we have included slope parameters η_f and η_r. At normal collector voltages, V_{bc} is a large negative value, so the second term in (2.21) is negligible. We then can ignore the second term and focus on the first one.

At low collector currents, $p = N_a$, where N_a is the acceptor concentration in the

2. For those of you who have been living alone under the polar ice cap for the last 20 years, SPICE is a time-domain linear/nonlinear circuit simulator developed at the University of California at Berkeley. It is the dominant product used in industry for designing integrated circuits.

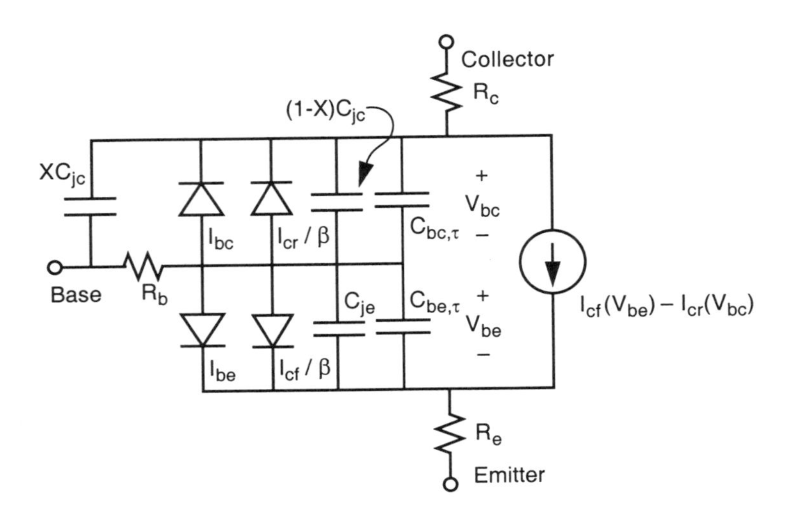

Figure 2.9 Gummel-Poon large-signal BJT model. In this model C_{jc}, the depletion component of C_{bc}, has been split into two parts, and extra diodes have been included to account for base-to-emitter and base-to-collector leakage at low voltages.

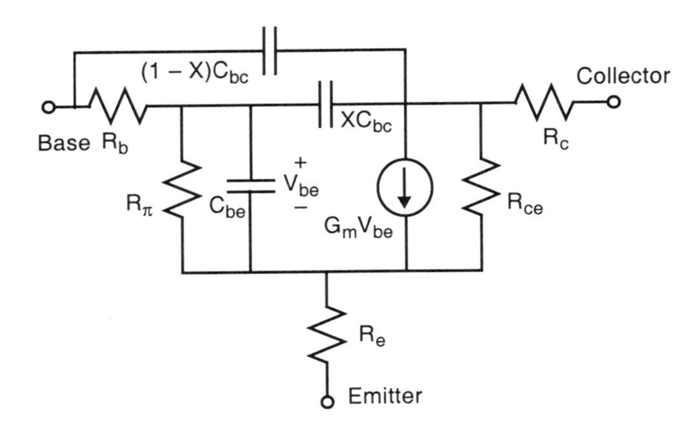

Figure 2.10 Small-signal equivalent circuit of a BJT, obtained by linearizing the circuit in Figure 2.9 in forward-active operation. In this mode C_{be} includes both depletion and diffusion capacitance, but C_{bc} has only a depletion component.

base, a constant. Since the first term is an exponential, a plot of $\log(I_c)$ versus V_{be} is a straight line. As current increases, however, and more holes are injected into the base, p begins to exceed N_a, and the $\log(I_c)$ versus V_{be} plot starts to flatten out.

Gummel and Poon showed that this phenomenon can be modeled by a curve with two asymptotes, one at low currents and one at high currents. Below the current $I_c = I_{kf}$, the slope is 58 η_f mV / decade of current, as predicted by (2.21) with $p = N_a$. Above this current, the asymptote has half this slope, 116 η_f mV / decade. The crossover point of the two asymptotes is I_{kf}, a parameter of the model.

This characteristic is especially important in large-signal circuits; without it, an analysis predicts greater output power than the transistor can, in reality, provide.

Base-Width Modulation

The width of the BJT's base-to-collector and base-to-emitter depletion regions are affected by changes in the collector voltage. This phenomenon, called *base-width modulation* or the *Early effect* (after J. M. Early [4]; this has nothing to do with timing), is manifested as an increase in collector current with collector-to-emitter voltage, a slope in the otherwise flat parts of the curves shown in Figure 2.6. This effect is described by a single parameter, called the *Early voltage*, V_A; the current is

$$I_c = I_{c0}\left(1 + \frac{V_{bc}}{V_A}\right) \tag{2.22}$$

where I_{c0} is the collector current as $V_{bc} \to 0$.

Low-Level Phenomena

Because of several effects, BJTs have disproportionately high base current at low collector currents. This phenomenon is modeled by extra diodes in parallel with the base-emitter and base-collector junctions. The leakage is relatively unimportant in power circuits but can be significant in small-signal circuits.

Figure 2.11 shows a plot of the base and collector currents, on a log scale, as a function of the base-to-emitter voltage, V_{be}. Because of this phenomenon, the dc current gain β, given by (2.6), is low at low collector currents. Because of high-level injection effects, β also decreases at high collector currents. Between these extremes there is a broad, intermediate range over which β is approximately constant.

Nonlinear Base Resistance

The base resistance of a BJT is nonlinear; it decreases as current increases, levelling off at high base current. Although not part of Gummel and Poon's paper, the equations for nonlinear base resistance often are implemented in circuit simulators and are loosely described as part of the Gummel-Poon model. The expression for base resistance is

$$R_b = R_{min} + 3(R_{max} - R_{min})\left[\frac{\tan(z) - z}{z\tan^2(z)}\right]$$
$$z = \frac{-1 + \sqrt{1 + 144 I_b / \pi^2 I_{rb}}}{(24/\pi^2)\sqrt{I_b / I_{rb}}} \qquad (2.23)$$

where R_b is the base resistance; R_{min} is the minimum base resistance, which occurs at maximum base current; and R_{max} is the maximum base resistance, which occurs at zero current. I_{rb} is the base current at which the resistance falls halfway to its minimum value. This phenomenon is important in BJTs, but because of their low base resistance, it may not be significant in HBTs.

This resistance should be viewed as a small-signal quantity, that is,

$$R_b = \frac{dV}{dI} \qquad (2.24)$$

where V is the voltage drop across the resistor at current I. Note that it is not correct to describe the voltage drop across this resistance as

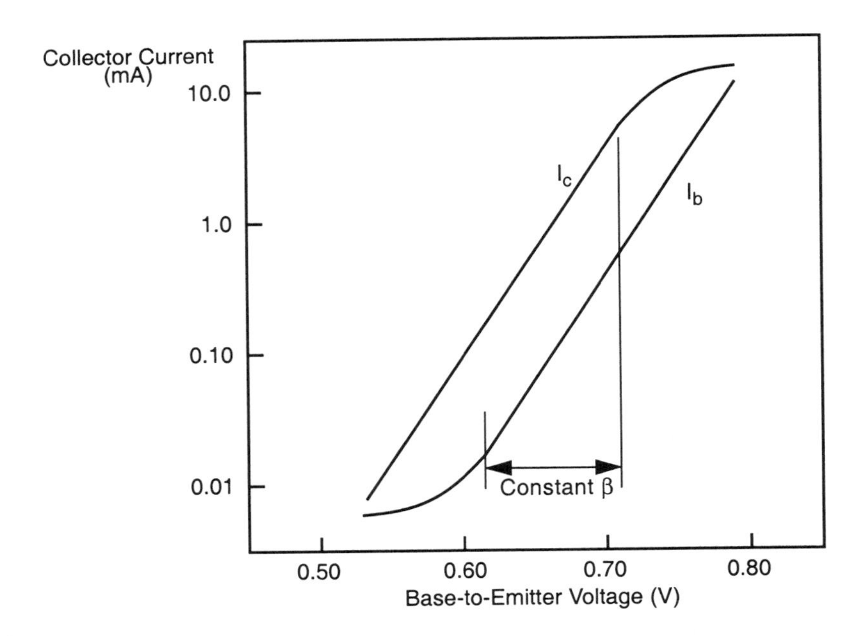

Figure 2.11 Collector and base currents, as a function of V_{be}, in a BJT. The dc current gain, β, is approximately constant over a broad range.

$$V = R_b(I) \cdot I \tag{2.25}$$

If you don't believe this, just differentiate (2.25)!

2.4 HBTS

Why are all BJTs made of silicon? Certainly, since mobility and saturated drift velocity are higher in GaAs than in silicon, it would seem logical to develop a GaAs BJT. Unfortunately, a number of practical problems, most associated with the limitations of p dopants, prevent the fabrication of diffused GaAs BJTs. Even if GaAs BJTs were made, the problem of high base resistance would persist; all p dopants of GaAs have poor mobility. Faced with these limitations, how do we exploit GaAs technology to make a high-performance bipolar device?

Advanced fabrication techniques, such as molecular-beam epitaxy (MBE), can be used to produce GaAs bipolars. These processes are expensive, however, and the improvement in performance must justify the cost. Thus, once we decide to use such processes, we must get as much of a return, in terms of performance, as possible. The epitaxial diffused device in Figure 2.5 is not a high-performance structure; it is largely a compromise between performance and cost. To justify our financial outlay, we must create something new.

2.4.1 Fundamental Properties

Figure 2.12 shows the cross section of an AlGaAs-GaAs[3] HBT. This mesa structure is typical of many types of HBTs. The top layer is the emitter. It consists of an n+ GaAs layer and several graded AlGaAs layers, which form the base-to-emitter junction. To minimize base resistance, the emitter is made as narrow as possible. To increase the emitter area, for higher peak current, a number of these cells can be connected in parallel.

The base is a thin layer of p GaAs. Because the high barrier is created by the heterojunction, not by doping, the base can be doped heavily to minimize base resistance. Even with a very thin base, an HBT's base resistance is quite low, often only a few ohms in small devices. Furthermore, because of the high base doping, HBTs generally do not suffer from high-level injection effects. Saturation of the $I_c(V_{be})$ curve is caused primarily by collector and emitter resistance.

The collector is conventional. The collector region is lightly doped n GaAs, and, as with the diffused device in Figure 2.5, an n+ region is used to minimize collector resistance and to create a uniform collector-current distribution.

HBTs can be used at frequencies well into the millimeter-wave region. Minimum noise figures are modest, typically a few decibels, but small-signal HBTs exhibit

---

3. If you like abstruse chemical notation, you'll love heterojunction devices. Things like AlGaAs-GaAs are about as simple as they come. Wait until we get to HEMTs!

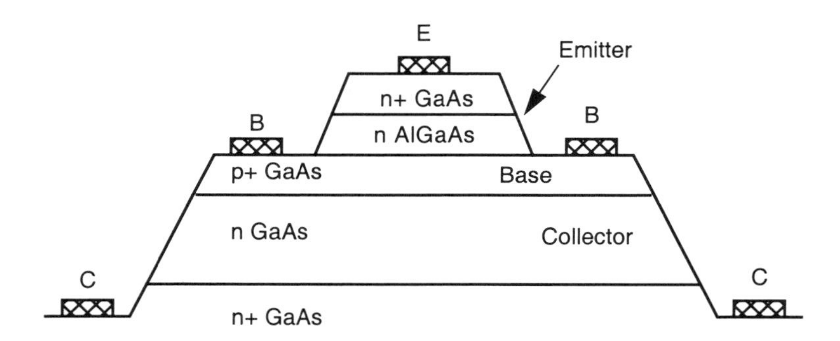

Figure 2.12 Cross-section diagram of an AlGaAs HBT. The mesa structure is universal. Usually several emitters are placed on a single mesa, and these cells are connected in parallel to create a larger device.

very low levels of intermodulation distortion (IMD). While the third-order intercept point (IP3) of a GaAs MESFET amplifier is usually about 10 dB above its 1-dB compression point, the IP3 of an HBT amplifier is often 20 dB or more above the compression point. HBT power amplifiers also exhibit low distortion, although not as dramatically low as small-signal amplifiers.

The urge to create a low-cost, high-performance HBT is, of course, irresistible. One response to this urge is the SiGe HBT. Such HBTs are realized in silicon technology and have very high performance (by silicon standards) and low cost (by GaAs standards). As such, they occupy a niche somewhere between conventional BJTs and GaAs HBTs.

2.4.2 Equivalent Circuit

Because HBTs are basically bipolar devices, their equivalent circuit and I/V and Q/V characteristics are similar to those of the homojunction BJT. There are a few differences, however, which we should note. The high base-doping densities of HBTs cause base-width modulation to be minor and high-injection effects to be virtually nonexistent. The low-current region in most types of HBTs is much wider than in BJTs; in fact, many HBTs have no constant-β region, and β increases monotonically with collector current. Base resistance is, of course, much lower, and only weakly nonlinear; base-resistance nonlinearity (Section 2.3.7) usually is negligible.

2.5 FET DEVICES

A wide variety of FET (*Field-Effect Transistor*) devices are in common use in the RF and microwave worlds. These include the following:

- Junction FETs (JFETs);
- Metal Oxide-Semiconductor FETs (MOSFETs);
- Metal Epitaxial-Semiconductor FETs (MESFETs);
- High Electron-Mobility Transistors (HEMTs).

We will consider these individually in the following sections.

All FETs have a few things in common. A FET has a conductive region, called the *channel*, which has *source* and *drain* contacts at its ends. Majority carriers move through the channel, from the source to the drain, reaching a velocity close to the saturated drift velocity. The number of carriers available is controlled by the electric field at the gate, which influences the channel in one of several ways. For example, the gate may form a *pn* or Schottky-barrier junction with the channel (JFET, MESFET) or may be separated from the channel by a thin insulator (MOSFET). By modulating the charge available for conduction, the gate voltage controls the channel current. Properly exploited, this leads to something called *amplification*.

The charge carriers in the FET's channel can be either electrons or holes. If the carriers are holes, the device is called a *p-channel* FET; if the carriers are electrons, it is an *n-channel* device. (Because *n* materials have higher mobility, RF and microwave devices are invariably *n*-channel, so henceforth we will consider only *n*-channel devices. The structures of *p*-channel devices are similar, and equations are identical for both types of FETs if all voltages and currents are appropriately reversed.) All FETs have a *pinch-off voltage*, the gate voltage at which the FET's channel turns off completely. Especially in MOSFETs, the pinch-off voltage may be called the *threshold voltage*.[4] If the pinch-off voltage is below zero in an *n*-channel device, the FET is a *depletion-mode* device; if above zero, it is an *enhancement-mode* device. JFETs, including MESFETs, invariably are depletion-mode devices; MOSFETs and HEMTs can operate in either enhancement mode or depletion mode.

2.6 JFETS

2.6.1 Structure and Operation

JFETs are the simplest FET devices and have the worst performance. They are quite adequate for many types of circuits at frequencies up to a few hundred MHz, however, and by understanding their limitations, we can understand the merits of other types of FETs.

Figure 2.13 shows a junction FET. It is fabricated on a low-conductivity silicon *p* substrate. An *n* channel is diffused into this substrate, along with *n*+ regions for ohmic contacts. Finally, a *p*+ gate is diffused into the *n* channel. The gate's longest

---

4. It gets worse than this. To circuit designers, the pinch-off voltage is the applied gate voltage that turns off the FET. To device designers, the pinch-off voltage is the voltage across the channel's depletion region (in the case of junction FETs) that equals the applied voltage plus the built-in voltage of the gate depletion region. Being circuit people, we use the former definition.

dimension is in the direction perpendicular to the plane of the figure; its shortest dimension is in the horizontal direction. Nevertheless, the former dimension is called the *gate width*, and the latter the *gate length*. The purpose of this counterintuitive terminology is to embarrass neophytes.

The p+ n gate-to-channel junction creates a depletion region in the channel. The width of this depletion region can be varied by the voltage applied between the gate and the source, changing the width of the undepleted region. A large-enough gate-to-source voltage will deplete the channel completely, preventing channel current. The voltage that does this is called the *pinch-off voltage*, usually a few volts.

A voltage applied between the drain and the source creates a current in the channel. If the drain-to-source voltage is low, the channel acts like a resistor, and changing its width simply changes the resistor's resistance. In this case, the FET is said to operate in its *linear*, or *voltage-controlled resistor*, region. At higher drain voltages, things get complicated. As the drain voltage increases, the channel current increases. At the same time, the voltage between the gate and the drain becomes greater than the voltage between the gate and the source, so the depletion region becomes narrower at the drain end. Because of this channel narrowing, the increase in channel current is not proportional to the increase in drain voltage, and the current starts to level off. Eventually, the drain end of the channel begins to pinch off, and the channel current cannot increase further. This occurs just as the drain-to-gate voltage reaches pinch-off. When this happens, the FET is said to be in its *current-saturated* (or simply *saturated*) *region*, also called the *pinch-off region*.

(The textbook explanation of this phenomenon is to say that, when the FET is pinched off at the drain end of the gate, no further increase in channel current is possible. The drain current then "saturates," that is, becomes constant with drain voltage. Thousands of college students have listened dutifully to this explanation without asking the obvious question, which probably has occurred to you right now: if the channel is really pinched off, how do the electrons get through? Of course, the channel really isn't pinched off. The electrons are squeezed into a narrow region between the depletion region and the bottom of the channel. The electrons' charge density in that region is very high, preventing further increases in the depletion width. Further increases in drain voltage do squeeze the region slightly narrower but build up more charge. The result is a stalemate in the competition between the depletion region width and the electron density, manifested as constant drain current.)

Other types of FETs operate in a generally similar manner, but often there are significant differences. In GaAs MESFETs, for example, the electrons in the channel reach saturated drift velocity long before the drain end of the channel pinches off, and this velocity saturation has a great effect on the performance of the device. Velocity saturation is present in Si JFETs and MOSFETs too, but it has a much less pronounced effect.

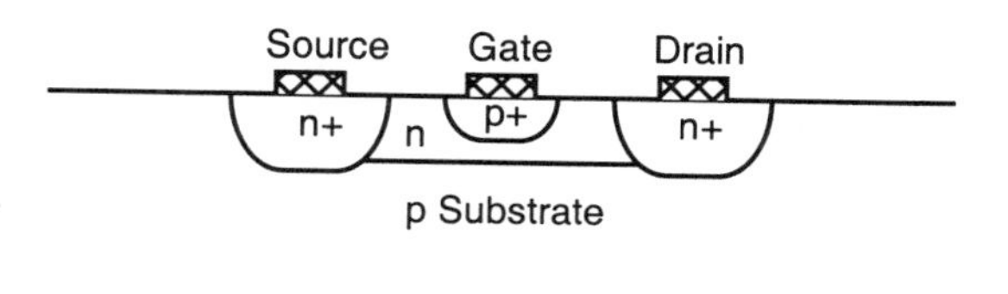

Figure 2.13 Cross section of a junction FET. The *n* region is the channel, and the *n*+ regions are ohmic contacts. The device is fabricated on a p substrate.

2.6.2 Electrical Characteristics

The I/V characteristic of a JFET is well described by the classical Shockley model. The drain current is

$$I_d(V_g, V_d) = G_0\left[V_d - \frac{2}{3}\left(\frac{(V_d + \phi - V_g)^{3/2} - (\phi - V_g)^{3/2}}{(\phi - V_p)^{1/2}}\right)\right] \tag{2.26}$$

where V_d is the drain-to-source voltage, V_g is the gate-to-source voltage, V_p is the gate pinch-off voltage (a negative quantity for *n*-channel FETs), ϕ is the built-in voltage of the gate-to-channel *pn* junction, and G_0 is a constant that has units of conductance. The voltages are internal quantities, that is, they do not include voltage drops across the drain and source resistances.

Equation (2.26) is nice to know but not very useful, since it is valid only if the gate-to-drain voltage, V_{gd}, is greater than the pinch-off voltage. This corresponds to linear operation of the device. JFETs are usually (but by no means exclusively) operated in saturation. When $V_{gd} \le V_p$ the device is saturated and the current is almost constant with increases in V_d. The drain current then is best expressed as

$$I_d(V_g, V_d) = I_{dss}\left(1 - \frac{V_g}{V_p}\right)^2 (1 + \lambda V_d) \tag{2.27}$$

where I_{dss} is the drain current at $V_g = 0$ and λ accounts for the FET's finite drain-to-source resistance. The parameter G_0 in (2.26) is chosen to make (2.26) and (2.27) equivalent at $V_{gd} = V_p$.

These expressions are adequate for most purposes, but they may not be adequate for very short-channel devices. In particular, the square-law I/V characteristic described by (2.27) probably should not be taken too literally. For example, (2.27) implies that JFETs have little or no third-order distortion. Although JFETs are very good in this respect, they are not *that* good.

One of the most important characteristics of a FET is its transconductance. High transconductance is necessary for achieving good gain in small-signal amplifiers. The transconductance is simply

$$G_m = \frac{dI_d}{dV_g} \tag{2.28}$$

evaluated in the saturation region.

Figure 2.14 shows a plot of a FET I/V characteristic, including the boundary between the linear and saturation regions, using V_g as a parameter. Note that, although V_g varies in equal steps, the curves are not equally spaced, and the transconductance ($\Delta I_d / \Delta V_g$) is highest at high values of V_g.

2.6.3 Large-Signal Equivalent Circuit

Figure 2.15 shows the large-signal equivalent circuit of a JFET. The following important parasitics are included: (1) source and drain resistances, the resistances of the ohmic contacts; (2) gate-to-source and gate-to-drain capacitances, which usually are treated as ordinary *pn*-junction capacitances; and (3) drain-to-source resistance, which is not shown explicitly but is part of the controlled current source. Although it is not shown, the *pn*-junction capacitance of the drain- and source-to-substrate capacitances sometimes are included.

2.6.4 Small-Signal Equivalent Circuit

As with the BJT, a small-signal equivalent circuit for the JFET can be generated by

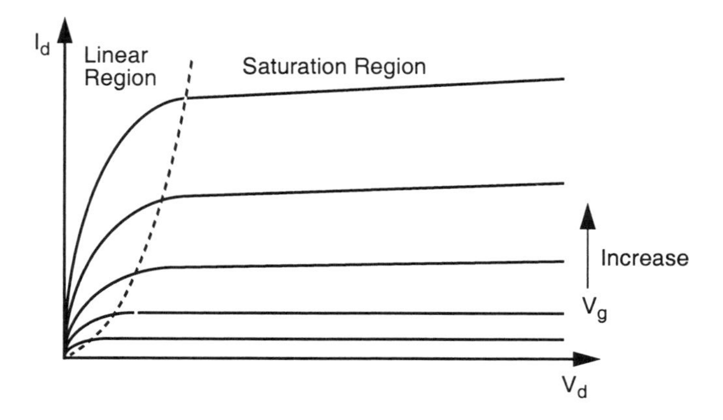

Figure 2.14 The JFET's drain I/V characteristic is divided into two regions: the current-saturation region and the linear region. Normally, the FET operates in current saturation.

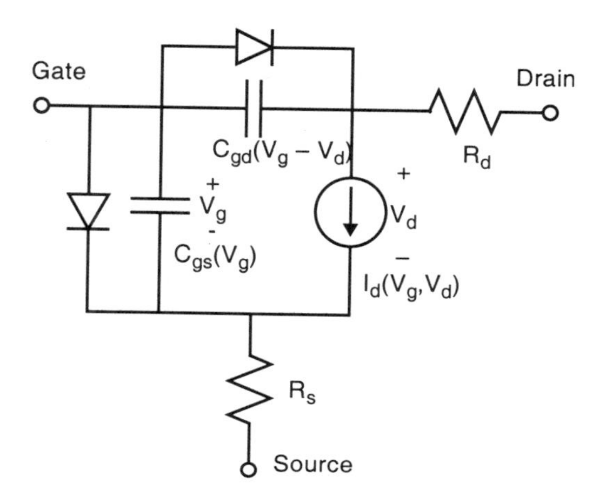

Figure 2.15 Large-signal equivalent circuit of a JFET. The gate-to-channel junction is modeled by two capacitances and two diodes. The resistance of the gate metal is not shown; in most cases it is negligible.

linearizing the large-signal circuit. Figure 2.16 shows such a circuit. It describes the FET in its saturation region only. The diodes of the large-signal circuit, which are normally reverse-biased in saturation, have been eliminated, and the controlled source replaced by its linearized form. R_{ds} is a new element; it is simply the linearized drain-voltage dependence of the large-signal source. Thus,

$$
\begin{aligned}
G_m &= \frac{dI_d}{dV_g} \\
G_{ds} &= \frac{dI_d}{dV_d} \\
R_{ds} &= 1/G_{ds}
\end{aligned}
\tag{2.29}
$$

2.6.5 Performance Characteristics

What's good about this, and what's bad? First the good: it's small and cheap and works pretty well for many purposes, especially if you don't like high frequencies. The rest of this section is reserved for the bad news.

In a FET, a small set of characteristics are critical to its performance. First, and perhaps most important, is the *transit time*, the time required for carriers to cross under the gate region. The shorter this is, the higher the device's transconductance and the greater its gain, especially at high frequencies. High transconductance is one

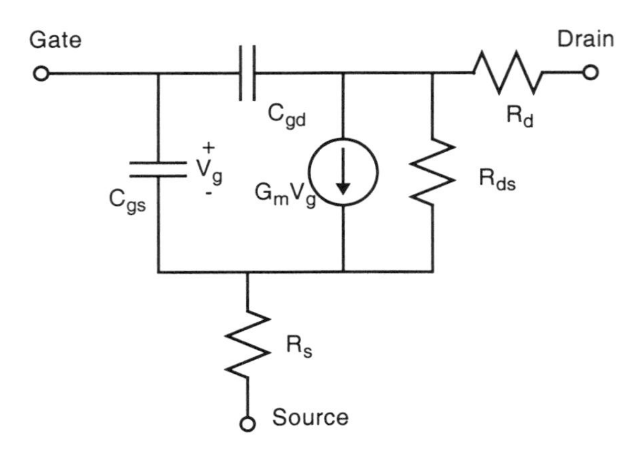

Figure 2.16 Small-signal JFET equivalent circuit. This is a linearization of the large-signal circuit.

of those things that invariably is good; as transconductance increases (and all else remains the same, of course), virtually everything about the FET gets better: gain, noise figure, and frequency response. The second characteristic is the gate-to-channel capacitance; again, the lower it is, the better. We noted that, in the saturation region, the drain voltage has little effect on the depletion region; only the gate-to-source voltage changes the depletion charge; therefore the gate-to-channel capacitance is essentially the gate-to-source capacitance. In the linear region, however, this is not the case, and the gate-to-channel capacitance is divided into the gate-to-source and gate-to-drain components.

The third characteristic is the material from which the device is made. Electrons in *n* silicon have lower mobility and saturated drift velocity than in GaAs or InP. As a result, parasitic resistances are higher (especially the source resistance, which affects gain strongly) and transconductance is lower.

Another problem with JFETs—in fact, with any diffused device—is that diffusion is a sloppy fabrication process. It is difficult to achieve precise dimensions and, especially, abrupt changes in doping density. Because of the large dimensional tolerances, a JFET's gate must be relatively long, especially in comparison with high-frequency devices made by other means, and its electron transit time is long as well. Poor high-frequency performance is the result.

The diffused *p* gate also has a relatively large surface area. Most of this surface does not contribute to the operation of the FET, but it does contribute gate-to-channel capacitance. Thus, its capacitance is high.

Finally, the *p* substrate has a number of effects. The use of a *pn* junction for the channel has important consequences. The junction creates a narrow second depletion region at the bottom of the JFET's channel, and the width of this depletion region depends on the channel-to-substrate voltage. In effect, the device acts as if there were a weak gate on the underside of the channel. This effect is called *backgating*.

Additional depletion regions around the source and drain contacts create parasitic capacitances. Even if such capacitances are small, they can be significant, since the JFET has a relatively high drain-to-source resistance. In integrated circuits, efforts are made to minimize parasitic effects, but they cannot be eliminated entirely.

2.7 GaAs MESFETS

How do we ameliorate the substantial performance limitations of Si JFETs? From the long list of deficiencies in Section 2.6, the requirements should be clear:

- Use a semiconductor material having higher mobility and higher saturated drift velocity than silicon.
- Employ more precise fabrication techniques.
- Avoid the use of a diffused *p* gate.
- Exchange the *p* substrate for a high-resistance one.

These are the characteristics of a GaAs MESFET.

A MESFET is a type of junction-gate FET that has several nice features:

- The gate is a Schottky (metal-to-semiconductor) junction. Avoiding diffusion processes in fabricating the gate provides a much shorter gate length, minimizing the gate-to-channel capacitance and the gate transit time;
- The FET's channel is an epitaxial layer grown on a high-resistivity substrate. This allows the channel dimensions to be controlled precisely;
- Although in the past silicon has been used for MESFETs, the most common material today is GaAs. GaAs has much higher mobility than silicon. Because InP has even mobility, MESFETS sometimes are fabricated on InP substrates as well. InP is much more expensive than GaAs, however, and InP technology is not as mature.

2.7.1 Structure and Operation

Figure 2.17 shows the cross-section of a modern MESFET. The device is fabricated on a high-resistivity substrate. To prevent impurities and crystal imperfections in the substrate from affecting the FET's channel, a relatively thick, high-resistivity buffer layer is first grown on the substrate. A thin and very pure epitaxial layer is used for the channel. The channel is grown relatively thick and etched thinner where the gate is located. This *recessed gate* has a number of advantages; the most important is minimization of the resistance between the source ohmic contact and the gate region.

In operation, the GaAs MESFET is in many ways similar to the Si JFET. There is one important difference, however, related to the way in which electrons move in GaAs. In both silicon and GaAs the electron drift velocity is proportional to electric field strength, and at high electric fields the electron velocity saturates. In silicon, the low-field mobility is moderate, and the onset of saturation is gradual. In GaAs, the low-field mobility is much greater, and the onset of saturation occurs at a much

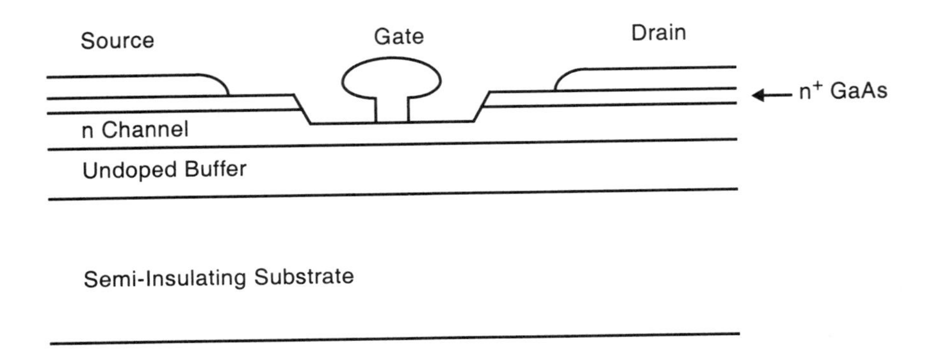

Figure 2.17 Cross section of a modern MESFET.

lower electric field strength. As a result, in GaAs devices saturation effects begin to occur well before the drain end of the channel pinches off, and current saturation occurs from a combination of electron velocity saturation and JFET-like pinch-off. One consequence is the formation of a much stronger charge dipole at the drain end of the channel in the MESFET. Because of this dipole layer, the Shockley model is not valid for MESFETs, and there is no accurate algebraic, physically meaningful model of such devices, as there is for JFETs. Thus, circuit designers usually depend on empirical expressions for modeling the I/V characteristics of GaAs MESFETS.

At microwave frequencies, even small parasitic resistances and capacitances can have a great effect on the operation of a MESFET. The resistance of the gate is one such parasitic. Although the gate is a metal and might be expected to have low resistance, it is actually a long, thin strip of very imperfect metal that easily can have a few ohms of resistance. Although a few ohms seems like a very small resistance, especially in comparison with BJTs and JFETs, it is indeed significant. In conjunction with the gate-to-source capacitance, it forms an R-C filter at the input of the FET, reducing its gain. By generating thermal noise, the gate resistance also degrades the noise figure of the device.

Two techniques are used to minimize gate resistance. The first is to reduce the length[5] of the gate strips. One way is to break the gate into a number of short strips connected in parallel. Another is to connect the gate pads to the gate strip at several points along its length. Although both techniques complicate the structure of the device and introduce additional parasitics, especially intermetallic capacitances, the overall effect is beneficial. A second technique is the use of a tee gate, a gate considerably wider at the top than at the semiconductor contact. Virtually all modern FETs use both techniques.

Another important parasitic is the resistance of the source's ohmic contact. This

5. In this discussion we view the gate as a strip of metal, apart from its use in the FET, so we employ the conventional meanings of *length* and *width*.

resistance causes two problems. First, by introducing negative feedback, it reduces the FET's gain. More importantly, because its thermal noise is part of the FET's input loop, it degrades the noise figure. To minimize source resistance, most FETs have large, heavily doped source regions. By minimizing the resistance between the source ohmic contact and the channel, the recessed-gate structure also helps to minimize the resistance in series with the source.

The inductance of the source contact is a significant parasitic in some types of circuits, especially power amplifiers. Surprisingly, the source inductance adds a *real* component to the input impedance of the device. The increase in impedance is approximately $G_m C_{gs} / L_s$, where G_m is the transconductance, C_{gs} is the gate-to-source capacitance, and L_s is the source inductance. This resistance is, of course, noiseless, but it reduces the gain of the device. For many reasons, power FETs are inherently low-gain devices, so any further gain degradation is a real problem. The universal solution is to use *via holes* (metal-filled holes through the substrate) to ground the sources.

Various parasitic capacitances limit the FET's maximum frequency of operation. The gate-to-source capacitance is primarily the Schottky capacitance of the gate-to-channel junction. Decreasing the channel doping decreases this capacitance but also decreases transconductance. Trading off transconductance against capacitance is a fundamental part of device design. The gate-to-drain capacitance is more interesting. In unsaturated operation (sometimes called *linear operation*), the Schottky capacitance is divided roughly equally between the gate-to-source and gate-to-drain capacitances. However, as the channel current reaches velocity saturation, the Schottky capacitance is represented entirely by the gate-to-source capacitance, and the gate-to-drain capacitance is almost entirely the intermetallic capacitance of the drain and gate metallizations. For this reason, and to increase gate-to-drain breakdown voltage, the gate often is located closer to the source than to the drain.

2.7.2 Device Size and Geometry

The length and the width of its gate is an important property of a FET. The shorter the gate, the better the performance in virtually all respects. However, making very short gates is difficult, requires expensive fabrication processes, and reduces manufacturing yield. Even so, for high-performance applications, gates as short as 0.1 μm can be fabricated. More mature technologies use gates of 0.25–0.5 μm length.

The wider the gate, the higher the FET's transconductance and gate-to-source capacitance and the lower its source resistance. These parameters are roughly proportional or inversely proportional to gate width. Increasing gate width also increases the FET's maximum drain current, so power devices invariably are wide devices. Increasing the gate width increases gate resistance, however, so power devices usually are broken into a large number of smaller sections connected in parallel. The gates of power devices may be several millimeters wide, and the structures of such large devices often are very complex.

High-frequency operation requires narrow devices with short gates; millimeter-wave devices may have gates as narrow as 25 μm. Millimeter-wave power devices usually use a large number of very short gate segments, all connected in parallel.

2.7.3 Electrical Characteristics

The electrical characteristics of MESFETs are qualitatively similar to those of JFETs. Both exhibit linear and current-saturated operation, although the boundary of these regions in MESFETs is not described as easily as in JFETs. In JFETs, current saturation occurs as the drain end of the channel pinches off. In MESFETs, current saturation occurs in part because of the narrowing of the drain end of the channel but also because of the onset of electron velocity saturation at relatively low drain voltages. The process is much more complex than in JFETs.

Except for the obvious observation that MESFET performance is far superior to JFET, it is difficult to compare the electrical performance characteristics of the devices; the devices are so different that it is hard to identify a valid criterion for comparison. We can make some general observations, however. For small-signal devices, JFETs have a transconductance of a few millisiemens and gate-to-source capacitance of about 1 pF; for MESFETs, these quantities are on the order of 100 mS and 0.25 pF. This implies that the frequency response of the MESFET is several hundred times that of the JFET, and indeed it is.

2.7.4 Large-Signal Equivalent Circuit

Figure 2.18 shows the large-signal equivalent circuit of a MESFET. In comparison to a JFET, it includes a few more circuit elements, not simply because the device is more complex but because of the high frequencies at which it operates. For example, the resistance of the JFET's "fat" gate usually is negligible, but the thin metal gate strip on a MESFET may have several ohms of resistance. When the MESFET is used as an amplifier, this resistance, in conjunction with the gate-to-source capacitance, creates an input RC filter in front of the FET. This has a strong deleterious effect on performance at high frequencies.

MESFETs have one characteristic that is inferior to that of a JFET: the drain-to-source resistance, R_{ds} in Figure 2.16. The small-signal JFET's R_{ds} is about 20 kΩ, and high-frequency measurements of this quantity agree well with dc measurements. In MESFETs, however, the high-frequency R_{ds} is only a few hundred ohms and does not agree at all with the dc measurement, which may be several kΩ . This difference is attributed to trapping effects in the MESFET's channel. These effects cause the low-frequency voltage gain of a MESFET often to be worse than that of a JFET or MOSFET. Although not terribly important in microwave applications, their low voltage gain limits the usefulness of MESFETs in many types of analog and digital circuits.

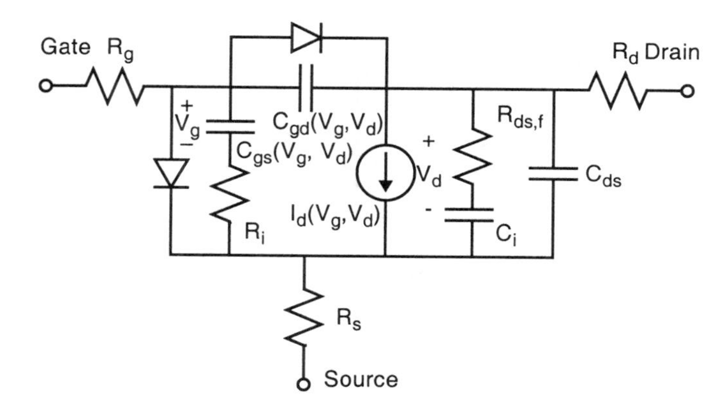

Figure 2.18 Large-signal equivalent circuit of the MESFET. $R_{ds,f}$ and C_i model the frequency-dependent drain-to-source resistance. R_i is called the *intrinsic resistance* or *gate-charging resistance* of the device. It accounts for additional channel resistance between the gate-depletion region and the source.

2.7.5 Small-Signal Equivalent Circuit

Figure 2.19 shows the small-signal equivalent circuit. As with the JFET circuit, it is a linearization of the large-signal circuit.

2.7.6 Performance Characteristics

The GaAs MESFET is the device most commonly used in industry for a wide variety of microwave circuits. MESFET ICs are used throughout the microwave region and lower millimeter-wave region. Although MESFET technology is more expensive than silicon, MESFET ICs still may be an alternative to silicon ICs where silicon performance is marginal.

At higher frequencies, especially in low-noise and millimeter-wave circuits, HEMTs may be preferable. HEMTs also have lower noise figures and higher maximum operating frequencies. On the other hand, MESFETs have higher current capability per gate width than HEMTs, so MESFETs usually are preferable for power amplifiers, especially below 30 GHz.

MESFETs have a number of additional idiosyncrasies that add variety to a circuit designer's life. (Many of these are shared by HEMTs; see Section 2.8.) We present a nonexhaustive list here:

- MESFETs achieve minimum noise figure at drain currents of approximately 10–25%[6] of I_{dss}, where I_{dss} is the drain current at zero gate voltage and normal

6. The politically correct figure is 15%. However, there is quite a bit of variability in this number.

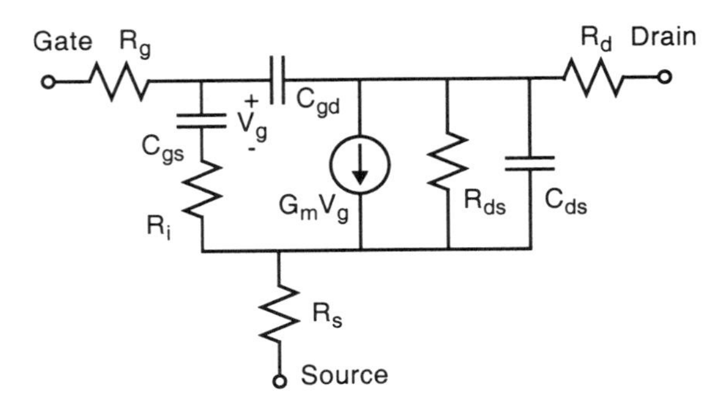

Figure 2.19 Small-signal MESFET equivalent circuit.

dc drain voltage. The lowest distortion occurs near 50% of I_{dss} or a little higher, and the highest small-signal gain near I_{dss}. Noise figure and distortion are quite sensitive to bias; gain is somewhat less sensitive. Thus, there is a significant trade-off between these characteristics.

- The source impedance that provides optimum noise figure is very different from the source impedance for best gain. Distortion is relatively insensitive to source impedance. Again, noise figure is quite sensitive to input tuning, but gain and distortion are less sensitive. The input VSWR of a low-noise MESFET amplifier invariably is quite high, and isolators or other means are necessary to achieve a good match.
- Like silicon FETs, MESFETs have a high dc gate-input impedance. Although the dc impedance is indeed high, MESFETs usually are operated at very high frequencies, where the reactance of the gate-to-source capacitance is low, and the resulting RF input impedance is low. Additionally, the resistive parasitics are small, giving the MESFET a low, high-Q input impedance. This makes it difficult to match over a broad bandwidth. Broadband FET amplifiers can be designed; however, unless isolators or quadrature-coupled stages are used (see Section 1.7), the input VSWR is always poor.
- The drain-to-source resistance (R_{ds}) of a MESFET is very different from that of a silicon device. Silicon FETs, both JFETs and MOSFETs, have very high R_{ds}. At dc, a MESFET's R_{ds} is moderate, perhaps a few thousand ohms, but at frequencies above a few MHz it drops rapidly to a few hundred ohms. One consequence is that the low-frequency voltage gain of GaAs MESFETs is lower than that of silicon MOSFETs. Thus, MESFETs are not well suited for use in analog ICs.
- Like most types of FETs, the gates of small-signal MESFETs have low breakdown voltages and are easily damaged by electrostatic discharge. Similarly, many types of electrical transients can "pop" a MESFET gate, and bias circuits must be carefully designed to prevent them.

- Similarly, it is imperative to turn on the gate bias before the drain bias in virtually all MESFETS. This practice prevents high drain current, which can damage the device. The time-tested procedure is to bring up the gate voltage gradually (don't just throw the switch!) and set it a little below pinch-off. Then, turn on the drain voltage, again increasing it slowly, and finally decrease the gate voltage until the drain current is the desired value. In some power devices, it is necessary to alternate between lowering the gate voltage and increasing the drain voltage, so the device does not self-destruct. Power supplies for FET circuits must be designed to turn on the FETs' gates before their drains, and to do it gradually.

2.8 HEMTS

Some people just can't be satisfied. You give them a FET with a noise figure of 2 dB at 20 GHz, and they want 1.5 dB at 30 GHz. Or better. For this reason, we have high-electron-mobility transistors, or HEMTs.[7]

In a HEMT, the electrons achieve much higher mobility than those in a MESFET; the result is higher transconductance, lower noise, and improved performance overall. This nice set of characteristics is achieved through the use of a heterojunction instead of a doped channel.

2.8.1 Structure and Operation

Figure 2.20 shows the cross-section of a HEMT. In many respects it looks like a MESFET, and the reasons for its structure are pretty much the same as for the MESFET. The difference, of course, is the use of a heterojunction instead of a simple doped channel.

The heterojunction consists of an *n* AlGaAs layer immediately under the gate, an undoped spacer layer, and an undoped GaAs layer. The spacer is very thin, on the order of a few tens of angstroms. The discontinuity in the band gaps of the AlGaAs and the GaAs causes a thin layer of electrons to form under the gate, at the interface between the undoped GaAs layer and the spacer. Because this layer is very thin and the electron concentration is very low, it is sometimes called a *two-dimensional electron gas*.

The electron concentration is modulated by the gate voltage, and in this way the gate voltage varies the drain current. Although the resulting transconductance may be very high, the amount of current available to produce output power is very low. To increase output power, multiple heterojunctions can be used.

Because the HEMT's structure is more complex than the MESFET's, there are more degrees of freedom for optimizing the device. One variation is the

---

7. Other acronyms are MODFET and TEGFET. I'm not sure what they stand for.

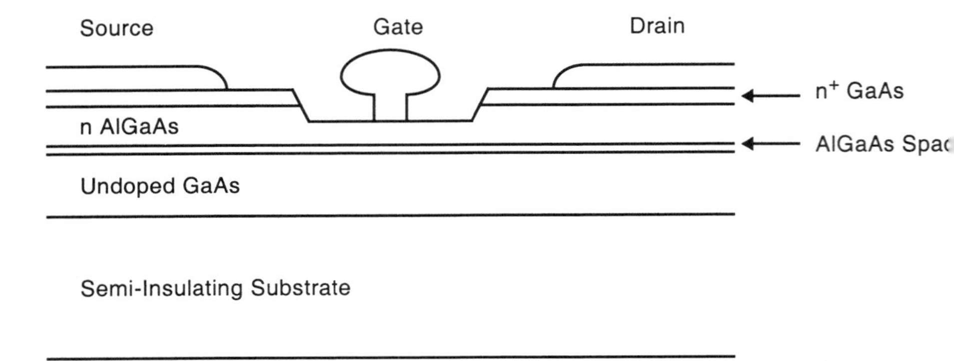

Figure 2.20 Cross section of a HEMT.

pseudomorphic HEMT, or *PHEMT*,[8] shown in Figure 2.21. A PHEMT uses an extra InGaAs layer between the *n* AlGaAs spacer and the GaAs layer; the resulting larger bandgap discontinuity creates more charge in the electron layer, increasing the transconductance and the output power. Unfortunately, this increase comes at the cost of a lattice mismatch between the three layers, which is taken up largely in the thin InGaAs layer. Because this layer is unnaturally compressed, it is called a *pseudomorphic layer*. The strain in this layer must be limited by limiting the amount of InAs in the structure, which in turn limits the electron charge density and the device's performance. Even so, PHEMTs exhibit significantly better gain, noise figure, and output power than simple AlGaAs HEMTs.

The HEMT's more complex structure brings us a number of opportunities for optimizing a device for a particular application. MESFETs have only two channel parameters to optimize: thickness and doping profile. With HEMTs we have a nearly infinite number of degrees of freedom for optimizing a device: the material characteristics and the dimensions of its many layers. We even can make multiple heterojunctions (which provide increased channel current in power devices) and optimize all the layers. The possibilities are endless.

2.8.2 Device Size and Geometry

The considerations regarding the size of MESFETs discussed in Section 2.7.2 apply to HEMTs as well. HEMTs are high-frequency, high-performance devices, so you are unlikely to find them with anything but the shortest possible gate lengths. Gate widths usually are optimized for millimeter-wave operation, but when costs decrease (as they always do), devices designed for high-performance, low-frequency use may become available.

---

8. Pronounced *pee'-hemt*, not *femt*. These things are important.

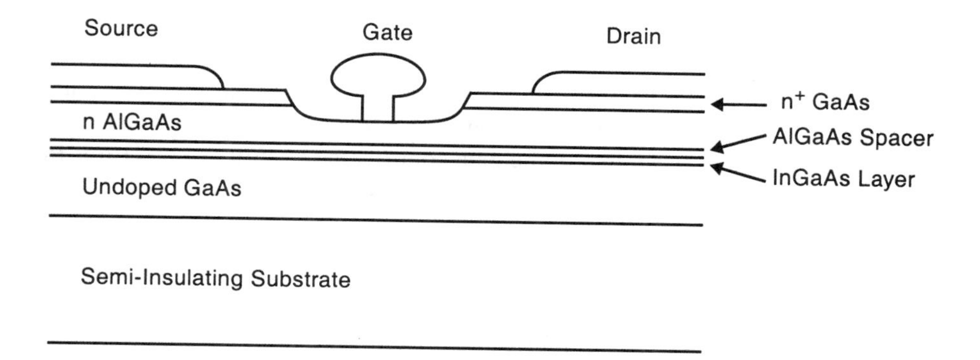

Figure 2.21 Cross section of a PHEMT.

2.8.3 Electrical Characteristics

HEMTs' drain I/V characteristics are qualitatively similar to MESFETs'. The greatest difference is probably the HEMT's lower dc output resistance.

The gate I/V characteristic of a HEMT is very different from a MESFET. The transconductance of a HEMT, as a function of gate voltage, usually has a distinct peak about halfway between pinch-off and maximum gate voltage. Furthermore, the pinch-off voltage can be positive, creating an enhancement-mode device. Although theoretically possible, enhancement-mode MESFETs rarely are produced.

2.8.4 Large-Signal Equivalent Circuit

The large-signal MESFET equivalent circuit of Figure 2.18 is applicable to HEMTs as well. Because the dc drain-to-source resistance is so low, often the $R_{ds,f}$ and C_i components are not needed. Most of the empirical expressions for $I_d(V_g, V_d)$ used for MESFETs are not applicable to HEMTs. HEMTs have their own empirical $I_d(V_g, V_d)$ functions.

2.8.5 Small-Signal Equivalent Circuit

The MESFET small-signal equivalent circuit, Figure 2.19, is applicable to HEMTs.

2.8.6 Performance Characteristics

The performance characteristics of MESFETs outlined in Section 2.7.6 are largely valid for HEMTs. Of course, HEMTs are much faster devices and have dramatically lower noise figures at very high frequencies. These characteristics make them useful as low-noise and power amplifiers at frequencies as high as 200 GHz. By the time you read this, maybe higher.

2.9 MOSFETs

Unlike JFETs and MESFETs, MOSFETs do not use either Schottky or *p-n* junctions for the gate. Instead, a MOSFET uses a metal gate separated from the channel by an oxide insulator. A voltage applied to the gate controls the quantity of electrons in the semiconductor under the oxide, and this electon layer becomes the channel.

2.9.1 Structure and Operation

MOSFETs come in *enhancement-mode* and *depletion-mode* varieties. Although both *p*- and *n*-channel devices can be fabricated, electron mobility is greater than hole mobility, so high-frequency transistors are exclusively *n*-channel. An *n*-channel depletion-mode device has a doped *n* channel, and the pinch-off voltage is negative. The gate voltage controls the depth of a depletion layer, much like the gate in a JFET. An enhancement-mode device has no doped channel, and its channel—an electron layer—is created when a sufficiently large positive voltage is applied to the gate. (The threshold voltage may be close to zero in some devices.) The more the gate voltage exceeds the threshold, the greater the electron density.

Figure 2.22 shows cross sections of enhancement-mode and depletion-mode devices. In both devices the oxide layer is made as thin as possible, consistent with fabrication limitations and the need for a reasonable breakdown voltage. The thinner the oxide, the higher the transconductance of the device. Thinning the oxide, of course, also increases the gate-to-channel capacitance; however, the overall effect is beneficial.

The operation of a depletion-mode MOSFET is similar to that of a JFET. The gate creates a depletion layer in the channel, and varying the gate voltage causes the depletion depth to change. The voltage that fully depletes the channel is the pinch-off voltage, and the device enters its saturation region when the channel is pinched off at the drain end, that is, the gate-to-drain voltage is less (more negative) than the pinch-off voltage.

The operation of an enhancement-mode device is somewhat different. An enhancement-mode MOSFET has no *n* channel; instead, there is *p* material under the gate. Because the drain-to-"channel" junction (the drain *n*+ region) is reverse-biased, there can be no drain current, beyond some slight leakage current. When a sufficiently large positive voltage is applied to the gate, however, an electron layer is generated under the gate, and conduction is possible. The voltage necessary to generate this layer, called the *threshold voltage*, is analogous to the pinch-off voltage in depletion-mode devices. As with the junction gate devices, when the gate-to-drain voltage drops below (is more negative than) threshold, the device enters its saturation region.

2.9.2 Device Size and Geometry

In junction FETs we are concerned primarily with gate length. Gate length is equally

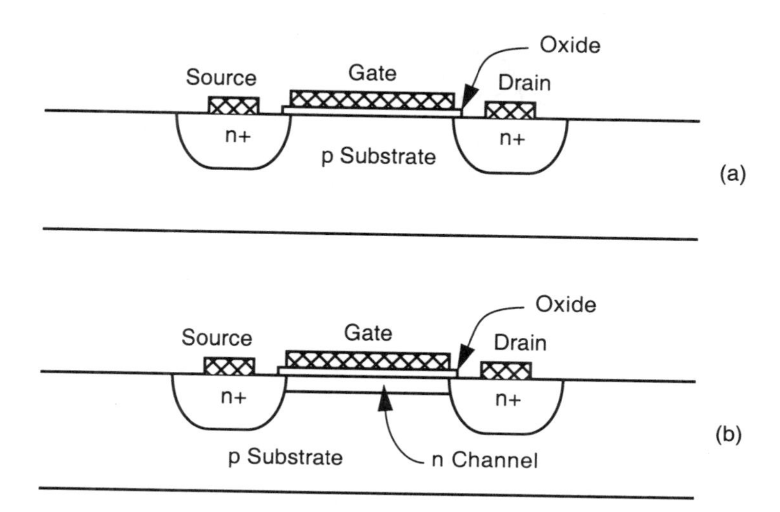

Figure 2.22 Cross section of (a) an enhancement-mode MOSFET and (b) a depletion-mode MOSFET.

important in MOSFETs, but oxide thickness is also a technology parameter critical to the device's performance. Modern fabrication techniques have produced silicon MOSFETs that are operable at frequencies of several gigahertz.

A variety of technologies exist for producing high-performance MOSFETs. Although most have been developed for digital applications, a few are useful for small-signal and power RF devices. One such technology is the laterally diffused device. An example of such a device is given in Reference [5] and shown in Figure 2.23.

2.9.3 Electrical Characteristics

The drain I/V characteristic is similar to that shown in Figure 2.14. For an enhancement-mode device, the current, in saturation, is given with reasonable accuracy by a simple expression,

$$I_d(V_g, V_d) = \frac{C_{ox}}{2}\frac{W}{L}(V_g - V_t)^2(1 + \lambda V_d) \tag{2.30}$$

where C_{ox} is the capacitance per area of the gate oxide layer, W is the gate width, and L is the gate length. V_t is the threshold voltage and, as with previous devices, V_g is gate-to-source voltage and V_d is drain-to-source voltage. It is interesting to note that (2.30) can be put into the same form as (2.27), the I/V expression for the JFET.

In the linear region (formally, $V_d < V_g - V_t$), the expression is

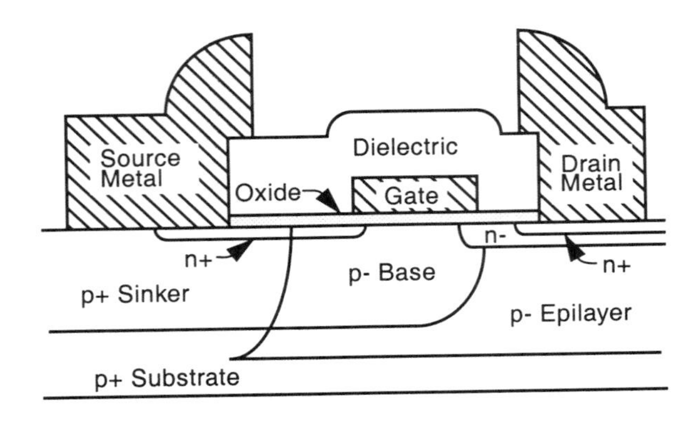

Figure 2.23 Cross section of a laterally diffused MOSFET. (After [5], used with permission.)

$$I_d(V_g, V_d) = C_{ox}\frac{W}{L}\left(V_g - V_t - \frac{V_d}{2}\right)V_d(1 + \lambda V_d) \quad (2.31)$$

which is somewhat different formally from the expression for the JFET, but not very different quantitatively.

Although (2.30) and (2.31) work well for ordinary MOSFETs, they are not particularly accurate for high-performance devices. Various types of short-gate effects and *backgating*, the effect of a substrate voltage on the channel, are not included in these expressions. Developing new MOSFET models currently is one of the favorite pastimes of academics, and many new MOSFET models have been published. One or two have been included in SPICE.

2.9.4 Large-Signal Equivalent Circuit

If the diodes are removed, the JFET equivalent circuit in Figure 2.15 is valid for MOSFETs. Models for more advanced, high-frequency devices may include more elements. For example, there often is a substrate-to-channel diode included to account for backgating (see Section 2.6.5).

2.9.5 Small-Signal Equivalent Circuit

The JFET equivalent circuit in Figure 2.16 is valid for MOSFETS as well as everything else (or so it seems).

2.9.6 Performance Characteristics

MOSFET performance is somewhere between that of a JFET and a MESFET, with a bit more weight on the JFET end. Many of the characteristics of JFETs discussed disdainfully in Section 2.6.5 apply to MOSFETs as well; one important exception is, of course, that the gates of MOSFETs generally are more precisely fabricated than the diffused gates of JFETs. They also can be made shorter, a small fraction of a micron in length. This makes MOSFET performance much better than that of a JFET, but still quite a bit below GaAs MESFETs. Still, the cost of MOSFETs is much lower than MESFETs, and at RF and lower microwave frequencies, the cost per performance, however defined, is about the same as for MESFETs.

REFERENCES

[1] Maas, S. A., *Nonlinear Microwave Circuits*, New York: IEEE Press, 1997.

[2] Ebers, J., and J. Moll, "Large-Signal Behavior of Junction Transistors," *Proc. IRE*, Vol. 42, 1954, p. 1761.

[3] Gummel, H. K., and H. C. Poon, "An Integral Charge-Control Model of Bipolar Transistors," Bell Syst. Tech. J., vol. 49, 1970, p. 827.

[4] Early, J. M., "Effects of Space-Charge Layer Widening in Junction Transistors," *Proc. IRE*, vol. 40, 1952, p. 1401.

[5] Camilleri, N., J. Costa, and D. Lovelace, "Silicon MOSFETs: The Microwave Device Technology for the 90s," *1993 IEEE MTT-S Int. Microwave Symp. Digest*, p. 545.

Chapter 3
Diode Mixers

Diode mixers are an example of a component that does nothing especially well but does most tasks at least adequately. Such components are the salt of the earth; technology could not exist without them. Diode mixers are low-cost, broadband components that exhibit moderate performance in terms of distortion, conversion loss, port isolation, spurious-response rejection, local-oscillator (LO) AM noise rejection, and low-frequency noise. Other types of mixers offer better performance for some of these characteristics, but none offers better performance for all of them.

Although mixers are used most commonly in communication receivers, they are also used in transmitters and signal generators. Mixers can be employed as modulators and phase detectors as well as frequency converters. The design of mixers for the latter applications is essentially the same as for receivers, except for occasional obvious differences. We examine frequency conversion in this chapter; in Chapter 5 we consider modulation.

Table 3.1 lists the types of mixers described in this chapter. It should be possible to find a diode mixer among those listed that is adequate for almost any ordinary application. For further general information on mixers, see [1–3].

3.1 DIODE MIXER THEORY AND OPERATION

3.1.1 Fundamentals

Frequency Mixing

Diode mixers are sometimes called *resistive mixers*, because mixing is caused by a time-varying resistance (or conductance). Consider a time-varying conductance having the waveform $g(t)$, where

$$g(t) = G_0 + G_1 \cos(\omega_p t) \tag{3.1}$$

Table 3.1 Mixers Described in This Chapter

Mixer Type	Characteristics	Typical Applications
Singly balanced, 180-degree ("rat-race") mixer	Approximately 15% RF and LO bandwidth. IF must be less than approximately 15% of the RF/LO frequency. Rejects (2,1) or (1,2) spurious response, but not both. Can be dc biased.	Not a general-purpose circuit. Best used for noncritical applications in integrated components. A poor choice when a broadband IF is needed.
Singly balanced 90-degree mixer, branch-line hybrid	Up to 20% bandwidth. LO- and RF-port bandwidths differ. No inherent IMD/spur rejection. Requires a good, broadband source VSWR over both the RF and LO bands at both ports, or imbalance occurs.	Not a good circuit; used more often than it deserves. Can be used for simple, noncritical applications.
Doubly balanced ring mixer; coupled-line baluns	Multioctave RF and LO bands. Narrowband IF, dc coupled. RF and LO bands can be widely separated.	General-purpose, broadband applications. The star mixer is a better biphase modulator.
Doubly-balanced ring mixer; "horseshoe" balun	Multioctave RF and LO bands. Broadband IF, dc coupled, which can partially overlap the RF and LO bands. RF and LO bands can be widely separated.	Most common type of general-purpose, commercial mixer. A very compact circuit.
Star mixer with Marchand baluns	Octave RF/LO band. Very broadband, dc-coupled IF but cannot overlap the RF/LO frequency range. RF and LO must cover the same frequency range.	Where the frequency plan allows it, this is a very good choice. High performance, good balance, broad bandwidth are easily achieved. Good biphase modulator or phase detector.

Note that the constant component of the conductance, G_0 must be greater than G_1, or the sinusoidal conductance waveform becomes negative over part of its cycle. We now apply a voltage v_s(t) to this conductance, where

$$v_s(t) = V_s \cos(\omega_s t) \tag{3.2}$$

a little trigonometry shows that the resulting current is

$$\begin{aligned} i(t) &= g(t)\, v_s(t) \\ &= G_0 V_s \cos(\omega_s t) + \frac{G_1 V_s}{2}\left[\cos((\omega_s - \omega_p)t) + \cos((\omega_s + \omega_p)t)\right] \end{aligned} \tag{3.3}$$

The time-varying resistance has generated *mixing products* at the difference frequency ω_s - ω_p and the sum frequency $\omega_s + \omega_p$. Usually the difference frequency is the desired output, but occasionally the sum frequency is desired.

In practical diode mixers the situation is not quite so simple. First, the conductance waveform is never a perfect sinusoid. In diode mixers it usually is a train of pulses that contains a large number of harmonics. Equation (3.1) becomes

$$g(t) = G_0 + G_1 \cos(\omega_p t) + G_2 \cos(2\omega_p t) + G_3 \cos(3\omega_p t) + \ldots \tag{3.4}$$

and we obtain mixing products between the RF and all harmonics of ω_p. Thus, the mixing frequencies ω_n, n = 0, ±1, ±2, ... are

$$\omega_n = |\omega_s + n\omega_p| \tag{3.5}$$

where n is any integer. A more common representation for (3.5) is

$$\omega_n = |\omega_0 + n\omega_p| \tag{3.6}$$

where ω_0 is the frequency $|\omega_s - \omega_p|$. This defines the same set of mixing frequencies as (3.5). ω_0 is called the *intermediate frequency*, or *IF*; ω_s is the *radio frequency*, or *RF*, and ω_p is the *local oscillator* frequency, or *LO* frequency. The spectrum of mixing frequencies is shown in Figure 3.1.

In deriving (3.3) we assumed that the voltage across the diode has only an RF (ω_s) component. In fact, the current circulates in the external circuit, generating voltage components at the same frequency as the current components. In general, the diode has both voltage and the current components at all mixing frequencies. These currents and voltages are "coupled" through the mixing process; changing the voltage or current at one mixing frequency affects those at another frequency. Changing the impedance terminating the diode at any mixing frequency therefore changes the voltage and current at all other frequencies.

In this sense, the time-varying conductance is a multiport network; however, in the diode, the ports are voltages and currents at the various mixing frequencies not

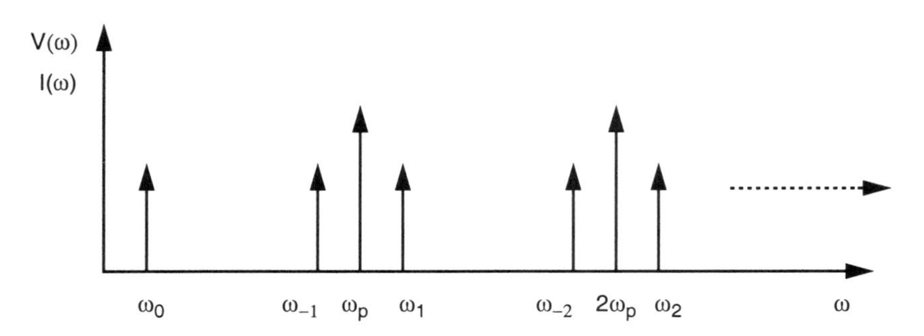

Figure 3.1 Small-signal mixing frequencies, ω_n, and LO harmonics, $n\omega_p$. The diode junction has voltage and current components at all these frequencies.

voltages and currents at physically separate ports (although, at least in theory, separate ports at each frequency could be created by means of filters). This brings us to a fundamental property of resistive mixers:

> The performance of a resistive mixer is determined entirely by (1) the shape of the conductance waveform and (2) the impedances terminating the diode (sometimes called *embedding impedances*) at all mixing frequencies.

This is starting to sound a little scary. It implies that, to design a mixer, we must design the circuits to have the proper source and load impedances at an infinite number of mixing frequencies! In practice, the problem is not so serious. First, diode mixers use, obviously, Schottky diodes. Above some frequency the diode's junction capacitance short-circuits the junction and the diode's termination becomes a short circuit. Fortunately, a short-circuit embedding impedance provides good performance. Second, as the LO harmonic number increases, the frequency component of the conductance becomes weaker, so it is less important in determining mixer performance. For these reasons it is customary to design the mixer for the proper RF and IF impedances and to take "pot luck" for the rest. Occasionally some effort is made to control the embedding impedances at other mixing frequencies, but, unless unusually high performance is needed, this casual approach is adequate.

LO Operation

There is one more detail we must examine. How do we obtain $g(t)$? The resistive junction of a Schottky diode has the I/V characteristic

$$I_j(V_j) = I_s\left[\exp\left(\frac{qV_j}{\eta KT}\right) - 1\right] \tag{3.7}$$

where (see Chapter 2) q is electron charge, K is Boltzmann's constant, T is absolute temperature, η is the ideality factor, and I_s is the reverse saturation current. The junction conductance is simply

$$g(V_j) = \frac{\partial I_j}{\partial V_j} = \frac{qI_s}{\eta KT}\exp\left(\frac{qV_j}{\eta KT}\right) \approx \frac{q}{\eta KT} I_j(V_j) \tag{3.8}$$

Therefore, if we know our junction-voltage waveform $V_j(t)$, we can substitute it into Equation (3.8) and obtain $g(t)$. Of course, we need to know $V_j(t)$ (which will approximate a rectified sine wave). That is what harmonic-balance analysis is for. It is important to remember that pumping the diode with the LO signal causes it to behave like a time-varying conductance. This time-varying conductance provides frequency mixing.

Matching

In order to minimize the LO power or to minimize conversion loss, we must conjugate-match the LO source and RF/IF port impedances to the diode. But what does a "conjugate match" mean in this case? Note that the diode generates current and voltage harmonics of the LO frequency, ω_p. The input impedance of the diode, which is the quantity that must be conjugate-matched, is

$$Z_i(\omega) = \frac{V_d(\omega)}{I_d(\omega)} \tag{3.9}$$

where $V_d(\omega)$ and $I_d(\omega)$ are the diode's terminal voltage and current, respectively, at the LO, RF, or IF frequency. Interestingly, although the diode is a nonlinear load for the LO and appears to be a time-varying conductance to the RF and IF, its input and output impedances at these frequencies are a single, time-invariant quantity.

Because the diode is nonlinear, these impedances vary with LO level; in general, all port impedances are high at low LO levels and decrease as LO level is increased. If the diode had no junction capacitance, and all ports were terminated in resistances, Z_i would always be real. In real circuits, however, the junction capacitance and reactive terminations at harmonics introduce a reactive part. Ideally, the diode should be chosen so that this reactance is small.

3.1.2 Other Important Performance Characteristics

Port VSWR

The input VSWR at each port is an important characteristic. Achieving a very low VSWR in a mixer, especially over a broad band, is difficult, but a VSWR of 2:1 or at worst 2.5:1 is a reasonable goal in almost all cases. In a balanced mixer (see Section 3.1.3) the LO port usually is easy to match; the RF port is more difficult. All ports are sensitive to LO power; the port impedance decreases as LO power increases.

Images

The mixer's IF is equal to the difference between the RF and the LO frequencies. Clearly there are two ways to obtain any IF frequency: the RF can be either above or below the LO frequency. Usually only one of these responses is desired; the other is called an *image*. The obvious problem with an image response is that it provides a path for interference. A less obvious problem is that it can increase the noise figure of a receiver by 3 dB. To prevent both problems, a mixer usually has an image-rejection filter at its input; image-rejection mixers also can be realized by a combination of balanced mixers and hybrids. See [1].

Intermodulation Distortion (IMD)

Mixers use Schottky-barrier diodes. Schottky-barrier diodes are nonlinear devices. Nonlinear devices cause distortion. The most common manifestation of nonlinear distortion is IMD, in which mixing products between harmonics of multiple RF signals are generated.

Most types of balanced mixers suppress IMD where the sum of the harmonics of the various RF excitations is an even number. The most effective way to reduce IMD in a diode mixer is to short-circuit the diodes at unwanted mixing frequencies and to use high LO power. FET resistive mixers, which we describe in Chapter 7, exhibit very low IMD.

Spurious Responses

One of the most important characteristics of a balanced mixer is its inherent rejection of certain even-order *spurious responses*. A spurious response is a mixing product between a harmonic of the RF and a harmonic of the LO. Most such products are very weak and fall outside the IF band. Occasionally, however, a "spur" falls inside the IF band and interference results.

Spurious responses are those that satisfy the relation

$$f_{IF} = mf_{RF} + nf_{LO} \tag{3.10}$$

where m and n are integers. By convention, this is called an (m, n) spurious response. Balanced mixers reject certain responses where m or n are even, and sometimes even-order intermodulation distortion. Be careful: not all types of balanced mixers reject all even-order responses! See [1] for further details on balanced mixers and their spurious-response properties.

3.1.3 Balanced Mixers

What does all this imply about the design of a practical mixer? It means that, in designing any mixer, we must do the following:

- Provide an LO signal to pump the diodes. Ideally, the LO port impedance will be matched to the diodes at the desired LO level.
- Provide coupling of the RF and IF to the diodes. Similarly, the RF and IF ports should be matched.
- Make sure that the RF, IF, and LO are isolated from each other. It is essential that the signal at one port does not leak out another port, that the port termination at the RF or IF does not affect the LO port, and that the LO termination does not affect the RF or IF ports.

The discussion in the Section 3.1.1 strongly implies that designing a diode mixer is a process of managing the diode's terminations at all mixing products and LO harmonics. Thus, we might expect filter and matching-circuit design to be a large part of the mixer-design process. Surprisingly, few practical mixers use matching circuits as such. Instead, the functions of these circuits are provided automatically by a balun, a hybrid, or a transformer, in combination with an array of two or four diodes. The result is a *balanced mixer*, the most common type of diode mixer used in practical systems. As well as providing a simple, elegant circuit that does not require complex matching circuits, balanced mixers provide a number of other benefits, including the rejection of AM noise from the LO, rejection of even-order IMD and spurious responses, and inherent port isolation.

3.1.4 Baluns and Hybrids

There appears to be quite a lot of confusion about the difference between a balun and a hybrid. This confusion stems in part from the fact that hybrids occasionally can be used as baluns (but baluns cannot be used as hybrids!). Here we try to set the record straight.

Hybrids

A microwave hybrid coupler or, more commonly, just *hybrid*, is a passive, lossless, reciprocal four-port component. If any one of those ports is excited, the signal is divided equally between two others and the fourth is isolated. Only two types of hybrids are possible. In one, the phase difference between the output signals is 180 degrees, or there is no phase difference, depending on the port excited. In the other, the outputs always differ in phase by 90 degrees. Hybrids are used as power dividers, as microwave adders and subtractors, and in various types of signal combining circuits. We described their properties in some detail in Section 1.7.1.

Hybrids are realized as transmission-line structures. Although many types of structures are possible, the circuits in this chapter will use only two: the "rat-race" 180-degree hybrid and the branch-line 90-degree hybrid. Other types of hybrids are the waveguide magic tee (180 degree) and the Lange coupler (90 degree). Ninety-degree hybrids can also be realized in stripline as broadside-coupled, 3-dB directional couplers. See [1,4] for further information on these types of hybrids.

Baluns

A balun is a transducer between a balanced and an unbalanced transmission line. The unbalanced port, of course, has a ground terminal, but the balanced port is "floating" in the sense that it has no ground terminal. Essentially, a balun converts an unbalanced input voltage, which is a combination of an even and an odd mode, into a purely odd mode at its balanced output. It is a type of mode transducer that converts the even-mode energy at the input into a purely odd mode. This definition, we realize, is considerably different from the one you will find in various trade-journal articles. (Naturally, we're right.)

A balun is a passive, lossless, reciprocal two-port component. Nevertheless, it can be treated as a three port, and often is. Since the output is an odd mode, we can use the two terminals of the balanced output as separate, unbalanced, out-of-phase ports. The balun then becomes a type of power divider, in which the output voltages are out of phase and equal in magnitude. The better a balun rejects the even mode, the better the phase and amplitude balance will be.

Why should we operate a balun in this manner, instead of simply with a floating load? For one thing, it's a good test to determine whether the balun really is eliminating the even mode effectively: if it is, the voltage split should be balanced within a fraction of one dB, and the phase difference at the output ports should be 180 degrees, plus or minus a few degrees. The even mode is responsible for such things as imperfect LO-to-RF port isolation in doubly balanced mixers. The better we do at getting rid of the even mode, the more we eliminate such undesirable phenomena.

A balun can be used as a simple power divider, but it's better to use a hybrid whenever possible. When a balun is used as a power divider, the three ports cannot be simultaneously matched; in fact, it can be shown theoretically that a lossless, passive three-port can never be matched simultaneously at all ports. Conventional power dividers have matched ports, so they usually are preferred, and the output ports are isolated. A balun might be preferable where a broadband 180-degree split is needed and the loads have low reflection coefficients.

The baluns we describe below are realized as interconnections of coupled transmission lines. Coupled lines are characterized by their even- and odd-mode characteristic impedances (Section 1.2.7). Most of these baluns are designed according to their odd-mode characteristic impedances, and the even-mode impedance is made as high as possible, ideally infinite. (This is, in part, how we eliminate the even mode: the even-mode energy approaches zero as the even-mode characteristic impedance approaches infinity.) Maximizing the even-mode impedance requires minimizing the capacitance between the coupled conductors and ground. Because of the air gap on both sides of its substrate, suspended-substrate stripline (Section 1.3.6) has low even-mode capacitance. SSSL baluns have very good performance; therefore, the baluns in many types of mixers are realized in this medium.

3.2 SINGLY BALANCED, 180-DEGREE "RAT-RACE" MIXER

3.2.1 Characteristics

The "rat-race" mixer is one of the most common types of mixers. It is easy to design and usually requires little effort to make it work. The greatest limitation is the rat-race hybrid, which has only about 15% bandwidth. A modification of the hybrid (described in Section 3.2.4) gives it greater bandwidth, but the structure may be difficult to fabricate on many types of microstrip substrates.

Because the hybrid has only 15% bandwidth, the RF and LO frequencies must be within about 15% of each other. This limits the IF to 15% of the RF/LO center frequency. Pushing the design to broader bandwidths or higher IF frequency inevitably results in poorer performance in virtually all respects.

The circuit usually requires a crossover in one of its microstrip lines.

3.2.2 Description

The circuit, shown in Figure 3.2, consists of a rat-race hybrid and two diodes. An IF filter is necessary, as is an IF-current return. No dc return is needed; the rectified dc LO current circulates in the two diodes.

The RF and LO are applied to a pair of mutually isolated ports; thus, both the LO and RF signals appear at the remaining two ports, to which the diodes are connected. Either port can be used for the RF or LO, but the spurious-response rejection depends on port selection: if the port marked "LO/RF" in Figure 3.2 is used for the LO, (2, 1) spurs are rejected but not (1, 2) [1]. The situation is reversed if the ports are reversed.

Depending on the choice of ports, one signal (RF or LO) is split with 180-degree phase difference, and the other is in-phase at the diodes. The resulting IF currents are in-phase in the diodes, which are simply connected in parallel at the IF. The IF filter grounds the diode terminal at the RF and LO frequencies; if the IF frequency is low enough, this filter may consist of only a simple capacitor or stub.

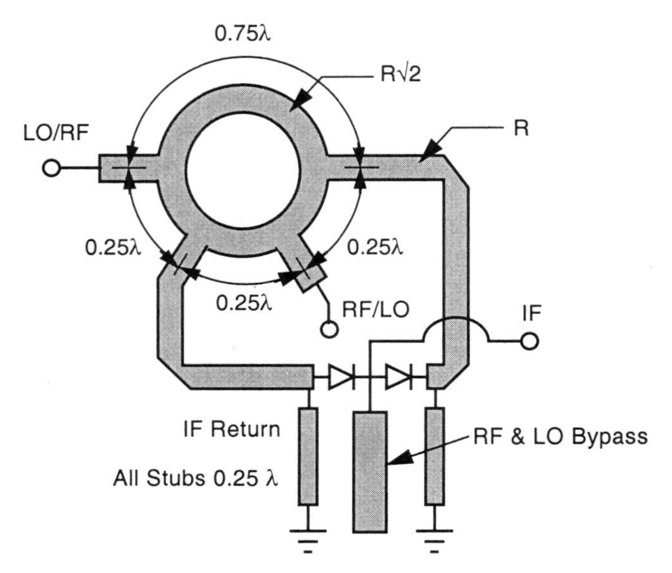

Figure 3.2 The rat-race mixer. R is the port impedance of the hybrid, usually 50Ω. The lines from the ring to the diodes must be equal in length. The stubs are 0.25λ at the center of the RF/LO band.

3.2.3 Design

Diode Selection

Generally, it is best to avoid the use of RF and LO matching circuits. Thus, to obtain a good VSWR, the diode's junction and parasitic capacitances must be negligible; if the RF/LO source impedance is 50Ω, the reactance of the total shunt capacitance must be at least 100Ω. Parasitic series inductance also must be negligible.

At frequencies below a few gigahertz, a low-cost epoxy-package diode may be adequate. At higher frequencies a beam-lead diode is more appropriate. A matched diode tee, consisting of two series-connected diodes in a single package, is a good choice.

Silicon diodes are available in low-, medium-, and high-barrier varieties. Low-barrier diodes provide good conversion loss at low LO levels, but higher IMD and spurious-response levels. Higher barriers require more LO power but have lower distortion.

Hybrid Design

Figure 3.2 shows all you need to know to design the hybrid. The characteristic impedance of the ring is $1.414R$, where R is the port impedance, usually 50Ω. The lengths of the sections are 0.25λ and 0.75λ, where λ is the wavelength at the center of the RF/LO band. The hybrid presents a 50Ω source impedance to each diode at the RF and LO; this is usually close to the optimum value.

At high frequencies the width of the microstrip lines at each port may be a significant fraction of the 0.25λ section of the ring. In this case it may be necessary to simulate the hybrid on the computer and to include a tee-junction model in the analysis.

RF/LO Design

If the diode's parasitics are negligible, as described above in *Diode Selection*, there is no need for matching circuits. The lines from the ports of the hybrid to the diodes should have equal lengths and characteristic impedances equal to the port impedance (usually 50Ω) and the diodes should be mounted as close together as possible. The open-circuit stub is necessary to ground the diodes at the center of the RF/LO band. For a 10% bandwidth, the stub's maximum impedance is

$$Z_{oc} \leq 0.1 Z_0 \tag{3.11}$$

where Z_0 is the characteristic impedance of the stub. Thus, a 50Ω stub easily covers a 10% band. Similarly, the IF-return stub should have a characteristic impedance high enough to make it "invisible" over the entire RF/LO band. For this short-circuit stub,

$$Z_{sc} \geq 10 Z_0 \tag{3.12}$$

In high-frequency mixers, where it may not have been possible to reduce the diode's parasitic capacitances to a negligible level, this stub can be used as an inductive tuning element to resonate the diode capacitance.

IF Design

Figure 3.3 shows the approximate IF equivalent circuit of the mixer. (It is assumed that the IF-return stub's impedance is much less than the port impedance of the hybrid.) The diode's output impedance at the IF, Z_d, is approximately 100-200 Ω, depending on the LO level.

This circuit should be analyzed on the computer to make sure that the IF bandwidth is adequate. If it is not, it may be necessary to increase the characteristic impedance of the short-circuit stubs or reduce that of the open-circuit stubs. This will affect the RF/LO bandwidth, of course, so this must be rechecked.

3.2.4 Variations

It is possible to increase the bandwidth of the hybrid to about 50% by replacing the 0.75λ section of transmission line with a quarter-wavelength pair of coupled lines, as

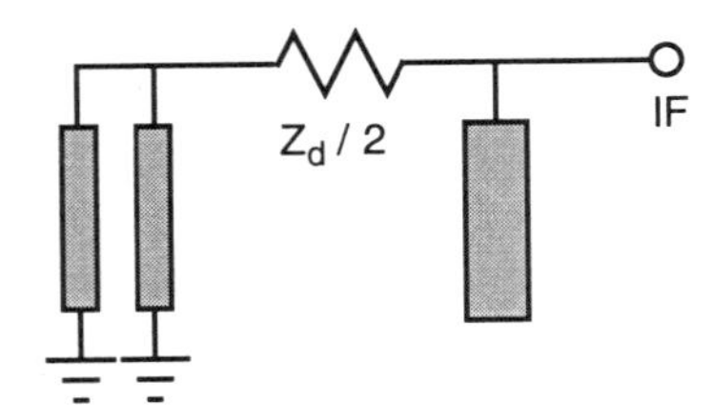

Figure 3.3 IF equivalent circuit of the rat-race mixer. The diode impedance at the IF, Z_d, should be approximately 100Ω, providing a good IF match.

shown in Figure 3.4 [5]. Unfortunately, the required even-mode impedance, 3.414R, is too high to be realized on many types of substrates. On such substrates the characteristic impedance of the ring can be reduced and quarter-wave transformers used at each port. For example, if the ring impedance is reduced to 50Ω, the port impedance becomes 35Ω, a value that is easily matched by a transformer. The coupled-line even- and odd-mode impedances are then 121Ω and 21Ω, respectively, the same values used in a Lange coupler.

Occasionally, the IF connection is made at the ring and the IF-port node in Figure 3.2 is grounded. (See [1] for an example.) This eliminates the stubs but requires IF blocking filters in the RF and LO ports. These filters sometimes are easier to implement, and the resulting layout may not require a transmission-line crossover.

It is possible to provide dc bias to the diodes. This is straightforward but requires several dc blocks and bias chokes. The use of dc bias allows an additional degree of freedom in adjusting the mixer and reduces LO-power requirements.

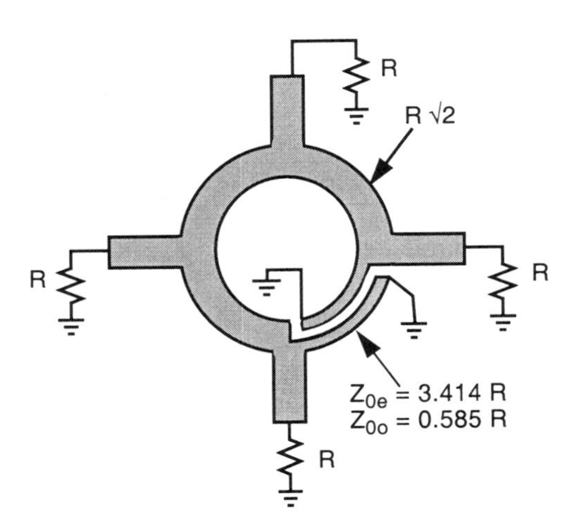

Figure 3.4 Broadband rat-race hybrid using a coupled-line section instead of the 0.75λ section in Figure 3.2.

3.2.5 Cautions

Don't try to tune a balanced mixer manually. If it is properly designed, there is little to gain but a lot to lose. Manual tuning usually upsets the balance of the circuit. This may not affect the conversion loss much, but it degrades all the beneficial characteristics of a balanced mixer: LO AM noise rejection, even-order spurious-response and intermodulation (IMD) rejection, and similar phenomena.

Similarly, if you use dc bias, be sure to retain symmetry in the circuit. This may require, for example, bias components at both diodes, even though they may be strictly necessary at only one. The reason, again, is to maintain balance in the circuit.

Nonlinear circuit simulation (by harmonic-balance analysis) is very effective in the design of rat-race mixers. In the simulation, adjust the LO level until the RF and IF VSWRs are minimum. It should not be difficult to achieve low VSWR at all ports at some LO power level. If the required LO level is too high, select a diode having a lower barrier height. If you still cannot achieve optimum port VSWR at the desired LO power, try replacing the lines from the hybrid to the diodes with quarter-wave transformers; a higher impedance at the diode will cause the conversion loss and VSWR to be optimum at lower LO power.

3.3 SINGLY BALANCED, 90-DEGREE MIXER

3.3.1 Characteristics

Because the branch-line hybrid most frequently used in its fabrication can be made circular, this type of 90-degree mixer sometimes is mistakenly called a *rat-race mixer*. In our opinion, that name really should be reserved for the mixer described in Section 3.2. This stuff is confusing enough without sloppy nomenclature!

This circuit, frankly, is not a good one. It offers only a few advantages over the rat-race mixer, and its disadvantages are significant (and largely unrecognized). We include it only because we realize that many people will build them anyway, and we want to help such benighted folk avoid the consequences of their folly. If this is not enough of a warning, read on.

The problems with this circuit are as follows:

- The signal paths in the branch-line hybrid from either the RF or LO port to the two diodes have different bandwidths.
- The mixer does not reject (2, 1) or (1, 2) spurious responses. (If you are not familiar with this concept, see [1].)
- It has relatively poor RF-to-LO and LO-to-RF isolation, which depends strongly on LO level. Isolation of 10 dB is typical.
- Proper operation requires that both the RF and LO ports have a good *source* VSWR over the entire RF and LO passbands. If this is not the case, the mixer will become unbalanced, and its already-minimal resemblance to a balanced mixer will be lost.

The advantages over the rat-race mixer are slightly greater RF/LO bandwidth and more convenient layout. Multisection branch-line hybrids can be used in place of the single-section hybrid we describe here, and these can have very wide bandwidths. For design information see [4].

3.3.2 Description

The circuit is shown in Figure 3.5. It is similar in structure to the 180-degree rat-race mixer; the only difference is in the use of a 90-degree branch-line hybrid instead of a 180-degree rat-race hybrid. The mixer is symmetrical; if the RF and LO ports are interchanged, no change in performance results.

At each diode, one signal—the RF or LO—is delayed by 90 degrees relative to the other. The IF currents are in phase in the diodes, and the IFs simply are connected in parallel. The IF circuit is identical to that of the rat-race mixer.

Although other types of quadrature hybrids can be used in this mixer, they seem to be unpopular, possibly because they are not as easy or as trouble-free to fabricate. A multistrip Lange coupler, similar to those used in amplifiers, could be used, for example, or a multisection branch-line hybrid. These hybrids would provide wider bandwidth, but the multisection branch-line hybrid would be relatively large and the Lange difficult to manufacture.

3.3.3 Design

Most aspects of the design of this mixer are identical to those of the rat-race mixer. The one difference is, of course, the hybrid. Its design is trivial; the hybrid consists of four microstrip sections, each one-quarter wavelength long, having the impedances R and $0.707R$, where R is the port impedance (invariably 50Ω) as shown in Figure 3.5.

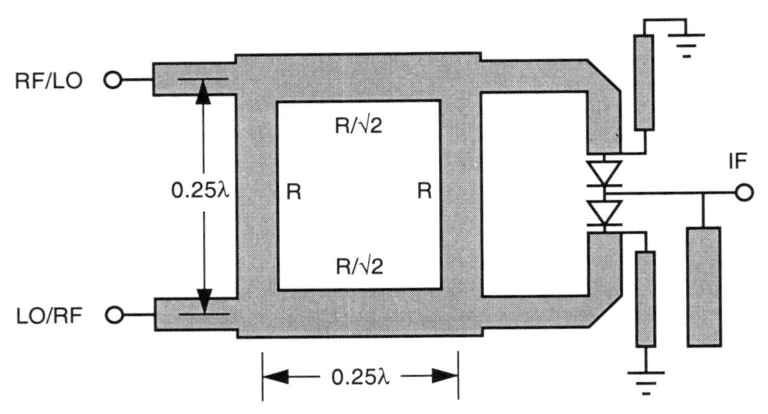

Figure 3.5 A 90-degree singly balanced mixer. The design and characteristics are similar to those of the rat-race mixer.

3.3.4 Variations

As with the rat-race mixer, dc bias can be provided to the diodes. The same considerations are valid.

An IF connection to the ring is possible but rarely used. This type of connection can be used to eliminate the IF stubs, but, except perhaps in special cases, no improvement in the circuit layout results.

3.3.5 Cautions

If we haven't frightened you away from this circuit by now, little else will do it. Again, the cautions in Section 3.2.5 are largely valid here. The problem most likely to attack both the neophyte and the experienced designer is probably the need for a low, broadband source VSWR at both the RF and LO ports, which is essential for proper operation.

3.4 DOUBLY BALANCED RING MIXER USING COUPLED-LINE BALUNS

3.4.1 Characteristics

This is an elemental yet practical circuit. It is a microwave realization of the classical transformer-coupled ring mixer. Unlike transformers, however, microwave baluns do not have a convenient structure, analogous to the transformer's center tap, to connect the IF port. The primitive IF circuit in this mixer provides a relatively narrowband (<1 GHz) IF. The horseshoe-balun mixer (described in Section 3.5) has much better IF bandwidth but somewhat less RF/LO bandwidth.

Because of the simple balun, the RF and LO bands in this mixer can be quite wide. Mixers having LO and RF frequency ranges of 2 to 26 GHz are not uncommon. Such a broad frequency range is probably beyond the capabilities of the horseshoe-balun mixer.

3.4.2 Description

The mixer is shown in Figure 3.6. The baluns, which must be realized on a suspended substrate, each consist of a pair of broadside-coupled striplines. The IF circuit consists of a pair of blocking capacitors and two high-impedance shorted stubs. The capacitors prevent IF leakage into the RF/LO port, and the stubs behave as low-value inductors at the IF.

The IF circuit admittedly is a little clumsy. Except at low frequencies, the capacitors do not provide a very effective IF block, and the inductance of the stubs is considerable. Less obvious is the fact that these components, in conjunction with the baluns, create spurious resonances that limit the IF bandwidth to about 1 GHz. The IF stubs also establish the lower end of the RF and LO bands. As frequency decreases, the stubs' impedances decrease, and they begin to short-circuit the RF and LO signals. Lengthening the stubs extends the RF and LO low-frequency response, but the increased inductance reduces the IF bandwidth.

3.4.3 Design

Diode Selection

Diode-selection considerations are much the same as those for the rat-race mixer (Section 3.2.3). Since there is no practical way to include a matching circuit, it is especially important to choose a diode having negligible parasitic capacitance and inductance. In fact, the upper frequency limit of the RF and LO band is established primarily by diode parasitics, especially the lead inductance and junction capacitance.

The use of a diode "quad" (four diodes in a single package) is essential. Individual devices simply are not practical in this type of mixer, if only because of layout difficulties. (If you don't believe this, try it!) Figure 3.6 shows an epoxy-packaged device, the type of diode most frequently used. The quad usually is mounted on its side, in an opening in the suspended substrate; this configuration minimizes lead length. Beam-lead quads also can be used, but plated-though holes must be used to connect the leads to conductors on the bottom of the substrate. The

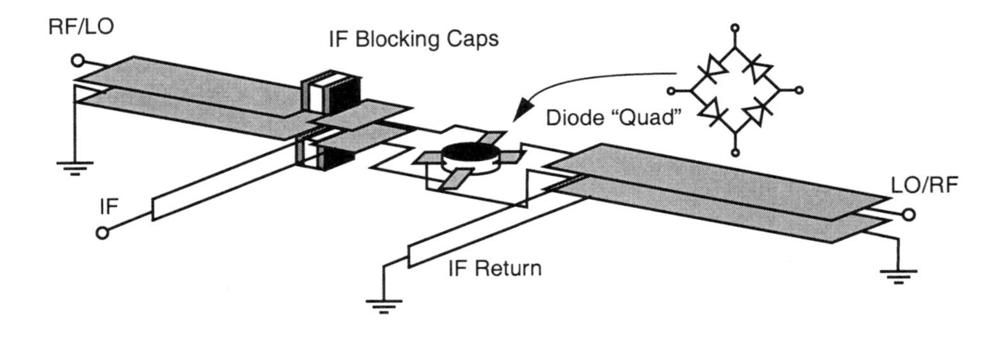

Figure 3.6 Doubly balanced diode mixer. The baluns consist of balanced, suspended-substrate transmission lines. Blocking capacitors are used to prevent IF leakage from the RF/LO port and a quarter-wave (at the center of the RF/LO band) stub provides the IF connection and current return. This simple IF circuit limits the IF band to approximately 0.5–1 GHz, but the RF and LO baluns can be very broadband. To minimize the lengths of its leads, the packaged diode "quad" often is mounted on its side in an opening in the substrate. Tapering the ground-plane side of the balun may increase the balun's bandwidth.

lengths of conductors necessary to connect the baluns to the beam-lead quad are surprisingly long, and the resulting inductive parasitics obviate much of the benefit of using a beam-lead diode quad.

Balun Design

The balun is designed around the odd-mode characteristic impedance of the coupled strips, Z_{0o}. If the even-mode characteristic impedance of the coupled lines, Z_{0e}, is high (as it should be), the coupled lines behave like a balanced transmission line having the characteristic impedance $Z_0 = 2Z_{0o}$.

For widest RF/LO bandwidth $Z_{0o} = 25\Omega$; this causes the pair of coupled lines to behave like a 50-Ω balanced transmission line, which is matched to the 50-Ω port impedance. This, of course, presents 50Ω to the diode quad. If a different impedance is desired, the balun's transmission line impedance can be modified so the balun operates as a quarter-wave transformer. However, its bandwidth will be considerably less.

The toughest trick in designing this type of balun is to manage the trade-off between size and low-frequency response. The lower the minimum RF/LO frequency, the longer the balun must be. However, the higher the even-mode characteristic impedance Z_{0e}, the shorter the balun can be, for a given minimum RF/LO frequency. Ideally, Z_{0e} should be at least $10Z_{0o}$. To achieve a high Z_{0e}, the air gap under and above the suspended substrate must be as wide as possible, and the strips must be as narrow as possible. However, to achieve the necessary odd-mode impedance, Z_{0o}, with narrow strips, the substrate must be very thin. For this reason it is not unusual to see composite substrates as thin as 5 mils (125 μm) used for these mixers.

The use of such thin substrates creates difficulties in fabrication. One possible alternative is to use thicker, high-dielectric-constant (high-ε_r) substrates. Unfortunately, that is not a good idea, because the difference in even- and odd-mode phase velocities in suspended, high-ε_r substrates degrades the performance of the balun. For this reason fiberglass-Teflon® composite substrates (Duroid® substrates) are used almost exclusively for this type of mixer.

It is common practice to taper the conductor on the ground-plane side of the substrate. This minimizes the generation of an even mode in the balun, which results in imbalance in the mixer. The underside conductor should be at least three times the width of the upper conductor at the input end. The shape of the taper doesn't seem to matter much, as long as it is gradual. Ideally, the taper should have a constant impedance, but maintaining a constant impedance is not easy and probably not worth the trouble.

This type of balun sometimes exhibits a passband "glitch[1]" at the frequency where the balun is one-quarter wavelength long (in terms of the odd mode.) The glitch is worst when Z_{0e} is not high enough. Tapering the ground-plane conductor helps to minimize this irritating characteristic.

We've been pretty qualitative so far. Now it's time to present some specific design rules:

- Use the thinnest composite substrate you dare. It should have a dielectric constant below 3.0.
- Select $Z_{0o} = 25\Omega$. If you have no software capable of analyzing SSSL, find the required width from microstrip tables. Use the width for a 25Ω microstrip line on a substrate half as thick as the one you are using.
- Determine Z_{0e}. You can use a stripline analysis to obtain an accurate value. Calculate the impedance of a stripline having the same ground-plane spacing, $\varepsilon_r = 1.0$, strip width equal to the width of your balun's upper strip, and strip thickness equal to your substrate's thickness. Then double it. This is Z_{0e}.
- The length is determined by the minimum RF/LO frequency and Z_{0e}. It is difficult to suggest something better than a rough estimate. One reasonable estimate is the following:

$$\theta = \operatorname{atan}\left(\frac{100}{Z_{0e}}\right) \tag{3.13}$$

where θ is the electrical length of the balun at the low end of the RF/LO band, calculated at the even-mode phase velocity.

RF/LO Design

Once the balun is designed, there isn't much left to do. First, note that the balun design described above results in a 50-Ω source impedance at the diode ring. But what impedance does each diode see? Fifty ohms, as it happens, because the balun output has two series connections of two diodes each. (We can ignore the effect of the second balun, because the terminals on the ring where it is connected are virtual ground points for the first balun.)

We can control the source impedance presented to the diodes by using the balun as a transformer. To do so, set $Z_{0o} = 0.5\,(Z_s\,Z_d)^{0.5}$, where Z_s is the source impedance (50Ω) and Z_d is the source impedance you want the diode to see. Of course, this reduces the bandwidth of the balun to that of the transformer. For more information of transformer bandwidth, see [4].

The IF stubs may limit the low-frequency performance of the balun. They usually are made one-quarter wavelength long at the RF/LO bandcenter, but the length can

1. Similarly sophisticated and equivalent technical terms are *suckout* and *VSWR spike*.

be adjusted considerably to ensure good low-frequency response. If the mixer exceeds one octave in bandwidth, the stubs will be one-half wavelength long within the RF/LO band and, in theory, should short-circuit the balun. As long as the stubs' impedances are very high, this does generally not happen, because at high frequencies the Q of a high-impedance stub is quite low.

IF Design

The four diodes are in parallel at the IF. As with the rat-race mixer, each has an output resistance of 100–200Ω (Section 3.2.3) so the IF output impedance is approximately 25–50Ω.

The IF equivalent circuit is shown in Figure 3.7. Except for the IF stubs, it is relatively straightforward. Each of the stubs should be treated as a single conductor; the two wires of the stub are close enough together that they should not be viewed as separate conductors. The characteristic impedance can be found from a number of sources, for example [6]. The capacitors, C_b in the figure, usually are quite small, less than 1 pF. They must be selected to have low impedance at the lower end of the RF/LO band but high impedance at the upper end of the IF band. If this is not possible, this type of mixer may not be appropriate.

Clearly, there are trade-offs between the IF and RF design. It may be necessary to go back and forth between the RF/LO and IF circuit designs a few times to get everything right. Analyzing the entire circuit on a harmonic-balance simulator allows such trades to be made in a straightforward manner.

3.4.4 Variations

Figure 3.8 shows a four-wire IF balun. This type of balun allows the IF to overlap the RF/LO band. Like the IF stubs in Figure 3.6, these stubs act as shorted quarter-wave stubs across the output of the RF and LO baluns. However, by interlacing the

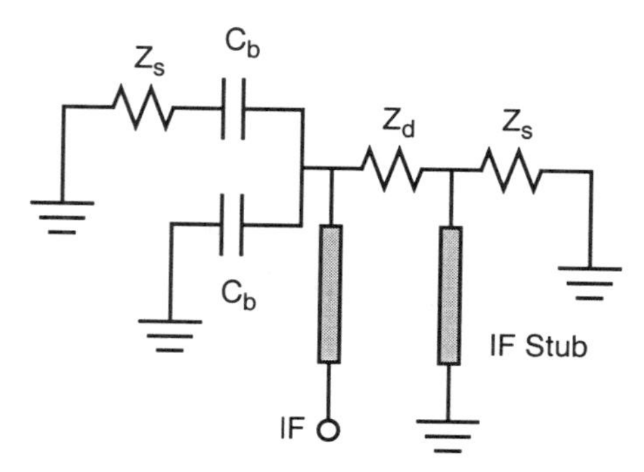

Figure 3.7 Approximate IF equivalent circuit of the doubly balanced mixer. C_b is the IF blocking capacitor, Z_s is the RF/LO port impedance (usually 50Ω), and Z_d is the diodes' output impedance at the IF, usually 25 to 50Ω.

conductors as shown, it is possible to make them approximate a transmission line at the IF. Determining the dimensions for a 50Ω line is a little tricky, but even if imperfect, it is an improvement over the arrangement in Figure 3.6.

Paradoxically, this does not work well if the IF does not overlap the RF/LO band. At low frequencies the IF-blocking capacitors are needed to prevent IF leakage from the RF/LO port. At IF frequencies within the RF/LO band, however, the baluns alone provide isolation and capacitors are not needed. In this case a very broadband overlapped IF is possible.

3.4.5 Cautions

In a nonlinear circuit simulator the conductors of the baluns and IF stubs should be modeled as coupled transmission lines. Avoid the temptation to model them as simple balanced transmission lines; this just doesn't work. For example, modeling the IF stubs as balanced transmission lines open-circuits the IF port! This happens because the stub carries the IF output in an even mode, but the even-mode impedance of an ideal transmission line is, theoretically, infinite. A transmission line model gives the RF and LO baluns infinite bandwidth and generally makes them seem to work too well.

If a tapered lower conductor is used in the balun, the circuit can be very sensitive to the quality of ground connections. This is especially important at high frequencies. Ground connections between the conductors and housing must be continuous and reliable, or strange behavior—housing resonances, passband "glitches" and poor port-to-port isolation—can occur.

Be careful of thin composite substrates. It is easy to damage them and to create shorts anywhere a conductor approaches the edge of a substrate. Transitions to coaxial connectors are especially vulnerable.

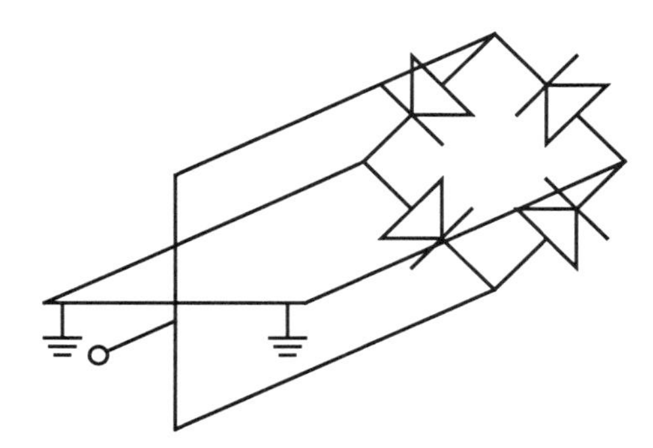

Figure 3.8 A four-wire IF balun. The conductors behave like stubs, as in Figure 3.6, for the RF and LO, but as a transmission line for the IF. To minimize LO-to-RF and LO-to-IF coupling through the balun, the conductors must be equally spaced.

3.5 DOUBLY BALANCED "HORSESHOE" BALUN MIXER

3.5.1 Characteristics

The "horseshoe" balun mixer is a variation on the doubly balanced mixer described in Section 3.4. Its LO uses a coupled-line balun, but the RF uses a structure we call a *horseshoe balun*, in recognition of its wide conductors and U-shaped structure. This balun provides an improved IF interface, but its RF bandwidth is not as wide as that of the coupled-line balun. The IF range can overlap the RF somewhat.

The main reason for using this structure is its broad, dc-coupled IF bandwidth: a bandwidth of several GHz is possible. The horseshoe section of the RF balun also provides additional even-mode rejection, resulting in improved balance and RF-to-LO isolation. The IF connection point of the horseshoe, which theoretically is a virtual ground for the RF and LO, is not one in practice. A stub may be needed to provide additional RF and LO rejection. This stub can be used to provide matching as well, improving the IF VSWR at high frequencies.

3.5.2 Description

The mixer is shown in Figure 3.9. It consists of a pair of coupled-line baluns, a diode quad, and a U-shaped coupled-line structure, the horseshoe. The horseshoe is simply two sets of coupled lines; its U (or sometimes V) shape is the only convenient way to lay out such a structure. The lines are broadside-coupled; one conductor of each set (for example, the inner horseshoe in Figure 3.9) is on the top side of the suspended substrate, and the other conductor (the outer horseshoe) is on the underside. The connection between the ground ends of this section is necessary in practice, although in theory it may seem extraneous.

As with the other mixer designs in this chapter, the horseshoe section of the balun is designed according to its odd-mode characteristic impedance; the even-mode impedance must be as high as possible. The odd-mode impedance must be relatively low, often in the range of 10 to 20Ω. This necessitates a wide line and a thin substrate. Because of the need to match even- and odd-mode phase velocities as closely as possible, the structure must be fabricated on a low-dielectric-constant substrate.

The LO balun is identical to the one used in the mixer described in Section 3.3. It provides part of the IF-return path; a stub is necessary to complete the IF-current return.

3.5.3 Design

The design of this mixer is essentially the same as for a mixer using coupled-line baluns (see Section 3.4). The design of the RF horseshoe balun is somewhat more complex, but it performs the same function as the simple coupled-line balun in the

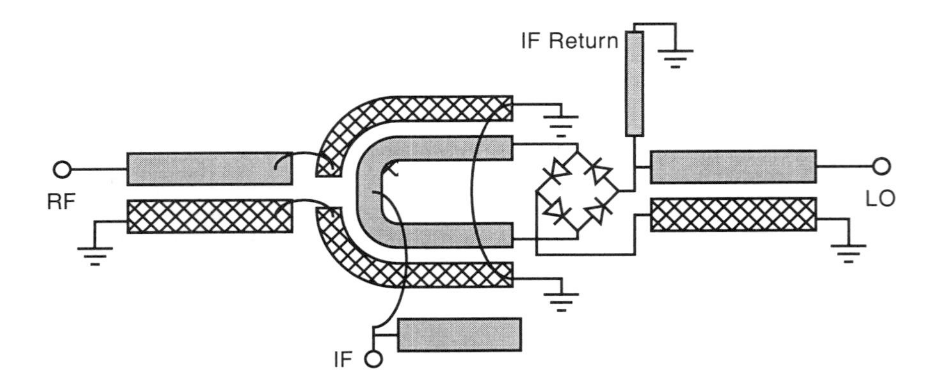

Figure 3.9 Doubly balanced ring mixer using coupled-line baluns and a U-shaped "horseshoe" balun. As with the circuit in Figure 3.6, the coupled-line baluns are broadside-coupled. The shaded sections are on top of the suspended substrate, and the cross-hatched ones are on the bottom. The open-circuit IF stub provides additional RF and LO rejection.

latter mixer. Specifically, the impedance presented to the diodes, within the RF band, is the same.

The length and impedance of the horseshoe section affect both the RF and IF frequency responses. A longer balun favors the RF, while a shorter one results in better IF bandwidth.

Diode Selection

Diode selection is essentially the same as for the other mixers in this chapter. See Sections 3.2.3 and 3.4.3 for a complete discussion.

Balun Design

The LO balun is identical to the one described in Section 3.4.3.

The RF equivalent circuit of the balun is shown schematically in Figure 3.10. Z_s is the RF source impedance, usually 50Ω, and $Z_{L,2}$ is the RF input impedance at the diodes (see Section 3.1). Since the ring consists of two parallel sets of two series-connected diodes, this is the same as the input impedance of a single diode. $Z_{0,o1}$ is the odd-mode characteristic impedance of the RF coupled line, and $Z_{0,o2}$ is the odd-mode impedance of the sections on each side of the horseshoe.

$Z_{L,2}$ is an intermediate impedance that is treated as a free parameter; it can be chosen arbitrarily by the designer. Obviously, certain values are better than others, so $Z_{L,2}$ should be chosen judiciously. Generally, broadest bandwidth results when $Z_{L,1} \cong Z_{L,2} \cong Z_s$. These impedances are given by the following relations:

$$Z_{0o1} = 0.5\sqrt{Z_s Z_{L,1}} \tag{3.14}$$

$$)_{o2} = 0.25\sqrt{Z_{L,1} Z_L} \tag{3.15}$$

which results in $Z_{0,o2} \cong 25\Omega$. This value is relatively low but can be achieved with broadside-coupled lines on thin, composite substrates.

The $Z_{0,o1}$ and $Z_{0,o2}$ sections are, ideally, one-quarter wavelength long at the center of the RF band. When the mixer is optimized on the computer, however, it may be necessary to modify these lengths. The horseshoe section usually must be shortened considerably, often to half the ideal value, to optimize both the RF and IF bandwidths. Shortening this section usually degrades the LO-to-IF and RF-to-IF isolation, so some care must be used in making this trade-off.

RF/LO Design

Harmonic-balance analysis is especially helpful in optimizing this circuit. The optimum balun lengths, as determined by nonlinear analysis of the complete mixer, are often quite different from those determined by linear analysis of the balun alone.

It is especially important to be careful in determining the length of the horseshoe. A shorter horseshoe has better IF bandwidth but may have restricted RF response at the low end of the band. The structure should be as short as possible without reducing RF bandwidth.

IF Design

The IF bandwidth of this mixer can be quite wide. The key is to manage the lengths of the horseshoe sections and the impedance of the IF return line.

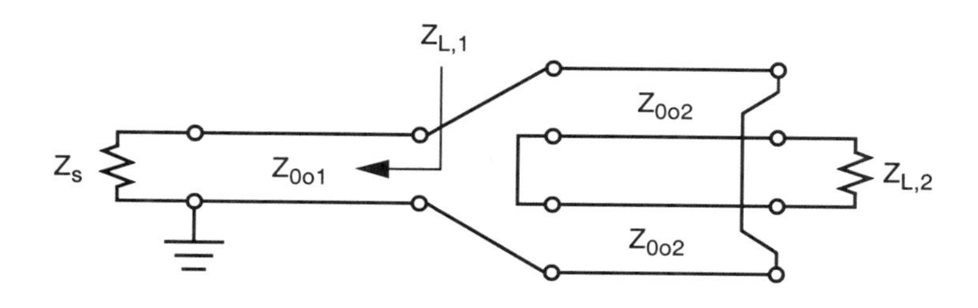

Figure 3.10 Equivalent circuit of the horseshoe RF balun.

As well as the trade-off of IF bandwidth with RF performance, discussed in *Balun Design*, there is a trade-off of the impedance of the IF return stub. The stub should be one-quarter wavelength long at the center of the IF band, and, to prevent degradation of the LO bandwidth, it should have a high characteristic impedance. At the IF frequency, however, the stub behaves as a series inductor, and if its impedance is too high, it can limit the IF bandwidth and introduce imbalance.

It often is necessary to include a stub in the IF circuit to improve the LO and RF rejection. This stub is one-quarter wavelength long at the RF/LO center frequency. It is capacitive at the IF, however, and its impedance can be selected to help match the IF port at the upper end of its range.

With careful design, it often is possible for the IF band to overlap part of the lower end of the RF and LO bands. For an IF that overlaps a substantial amount of the RF or LO bands, the four-wire IF balun (see Figure 3.8) is more appropriate.

3.5.4 Variations

Occasionally, the RF and LO ports are reversed. This results in wider RF bandwidth but narrower LO bandwidth. The LO-to-IF isolation also may be worse.

Other types of baluns, such as the Marchand balun described in Section 3.6.2, can be used at the RF and LO ports instead of simple, coupled-line baluns. These may provide better performance in some respects.

3.5.5 Cautions

The design of this mixer is somewhat more complex than that of the other mixers, but if the various design trade-offs are handled well, remarkably good performance is possible. The trickiest part is the adjustment of the horseshoe length. It is essential to monitor all important parameters of the design while this length is adjusted; it is possible to obtain a design that is very good in some respects (for example, RF VSWR and conversion loss) while making it unacceptable in other respects (for example, RF-to-IF isolation). Nonlinear analysis is almost essential in this process.

Some of the traditional fabrication techniques for this type of mixer are pretty bizarre. The strangest trick is to slit the thin, composite substrate around the RF balun, twist the balun, and solder its ends to the gap in the outer horseshoe. This structure is expensive fabricate and unreliable in use. As with the coupled-line-balun mixer, the epoxy-packaged diode quad is often mounted on its side in a hole in the substrate. The leads are then soldered to the striplines. The resulting parasitics are fairly large.

In both linear and nonlinear analyses of these baluns, it is essential to include the even-mode characteristic impedances of the baluns' coupled lines. Do not make these impedances very large, in an attempt to have an ideal circuit, or use transmission lines instead of coupled lines. This effectively open-circuits the IF.

3.6 DOUBLY BALANCED STAR MIXER

3.6.1 Characteristics

The star mixer is a very nice circuit for many applications. It provides good performance over octave bandwidths and a broadband, low-VSWR IF. All port-to-port isolations are excellent. Its primary limitations, however, are (1) the IF cannot overlap the RF or LO bands, and (2) the RF and LO ports must cover the same bands; unlike the ring mixers, the RF and LO baluns cannot have different frequency ranges.

3.6.2 Description

Figure 3.11 shows the mixer. The LO and RF baluns are oriented at right angles, and each consists of an input line (on the top of the suspended substrate) and a set of lines (on the underside of the substrate) to which it is electromagnetically coupled. The lines are one-half wavelength long at the center of the RF/LO band. The diodes are connected in a cross configuration (more elegantly called a *star*) in the center of the structure. The IF is connected to the diodes' common node, and the coupled lines of the balun provide the IF return.

Understanding the coupled lines is the key to understanding the operation of the mixer. We start by considering a single balun (let's say it's the RF). The parallel conductors on the underside of the substrate have the same voltage along them, so the gap between them has little effect on the balun. Its only effect is to reduce the coupling between the lower and upper conductors slightly, so to a good approximation we can model it as shown in Figure 3.12(a). This is a realization of a classical structure called a *Marchand balun*. Splitting the coupled conductors simply provides two outputs, as shown in Figure 3.12(b); these outputs are in phase, and each has twice the impedance of the single output.

Now, what happens when we connect the LO balun? Clearly, the LO voltage excites the RF balun in such a way that LO energy does not couple to the RF port; isolation is ensured. The RF balun looks like nothing more than a pair of shorted, quarter-wavelength stubs, which are open circuits at their inputs. Thus, as long as the impedances of those stubs is high, the RF balun is invisible to the LO. Similarly, the LO is invisible to the RF. Both, however, apply voltage to the diodes in the correct phase for a doubly balanced mixer.

This type of balun has a couple of subtle, yet important, characteristics. First, remember that the RF balun of a ring mixer is connected to points in the diode ring that are LO virtual grounds, and the LO balun similarly is connected to RF virtual ground points. This completely decouples the RF and LO baluns, allowing them to have different passbands. In the star mixer, however, the baluns are connected directly, and their decoupling requires that both baluns have the same center frequency. Each balun must cover *both* the RF and LO bands. Don't try to make the RF and LO baluns different lengths; this idea just doesn't work!

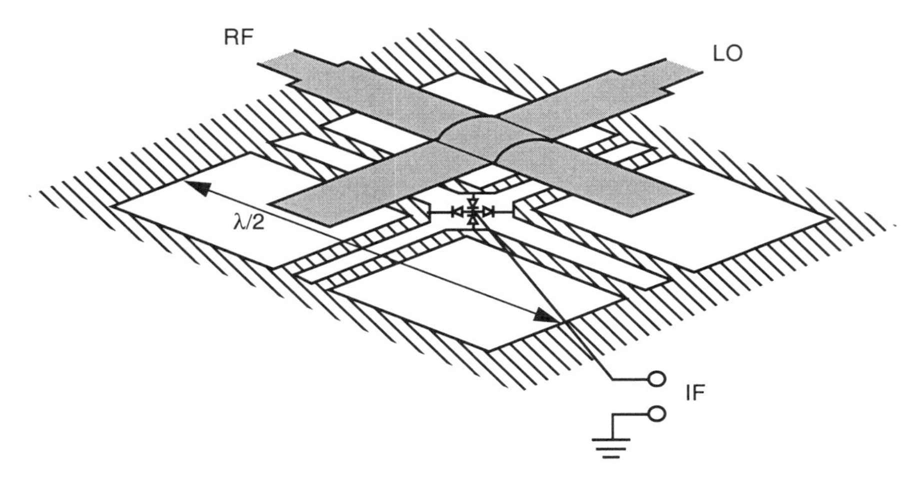

Figure 3.11 The star mixer. The RF and LO strips are coupled electromagnetically to conductors on the underside of the substrate. The orthogonal arrangement necessitates an undesirable crossover. The IF is connected to the center of the "star" of diodes, on the underside of the substrate. The low parasitic inductance of this connection results in a broad IF bandwidth.

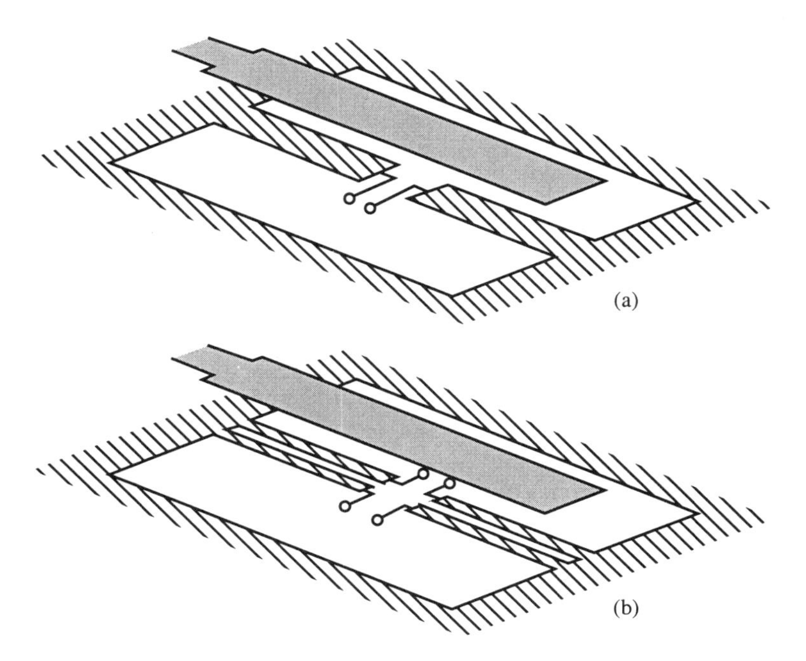

Figure 3.12 The evolution of the star-mixer balun. A Marchand balun (a) is modified to provide two outputs. Two of these baluns are connected to form the star-mixer balun.

Second, the IF currents excite both baluns in an even mode. The even-mode output impedance of a Marchand balun, within its bandwidth, is very high, so the IF cannot overlap the RF/LO band. In practice, the upper end of the IF band must be no greater than about 80% of the lower end of the RF/LO. However, outside the RF/LO band, the IF parasitic inductance is very low, and a broadband, low-VSWR IF can be achieved easily.

3.6.3 Design

This mixer's design is a little different from the previous ones. The same principles apply, however, and the same assumptions about the RF/LO input impedances at the diodes are valid.

Diode Selection

The standard philosophy applies: diodes are best chosen according to the information in Sections 3.2.3 and 3.4.3. Don't scour your diode catalogs looking for a star-mixer quad; they don't exist. Use two diode tees instead.

Balun Design

The simplified balun in Figure 3.12(a) can be used as a prototype for the design of the balun. As with other coupled-line baluns, the even-mode impedance should be as high as possible. The odd-mode impedance is

$$Z_{0o} = 0.5\sqrt{Z_s Z_L} \tag{3.16}$$

where Z_s is the source impedance (50Ω) and Z_L is the RF or LO impedance looking into a single diode, usually in the range of 50 to 100Ω. The balun is one-half wavelength long at the center of the RF/LO band. Unless Z_L is unusually high the bandwidth will be approximately 2:1.

RF/LO Design

The tricky part of the design of this mixer is the model for the three-conductor balun. An acceptable model is the parallel combination of coupled lines shown in Figure 3.13. Unfortunately, this model does not account for the gap between the two coupled lines, but, fortunately, the mixer's performance is relatively insensitive to the width of this gap. A good general rule is to make the gap approximately one-quarter of the width of the prototype balun's conductors. The even-mode and odd-mode impedances should be twice those of the prototype balun of Figure 3.12 (a), not the even-mode and odd-mode impedances of the individual coupled lines in Figure 3.13.

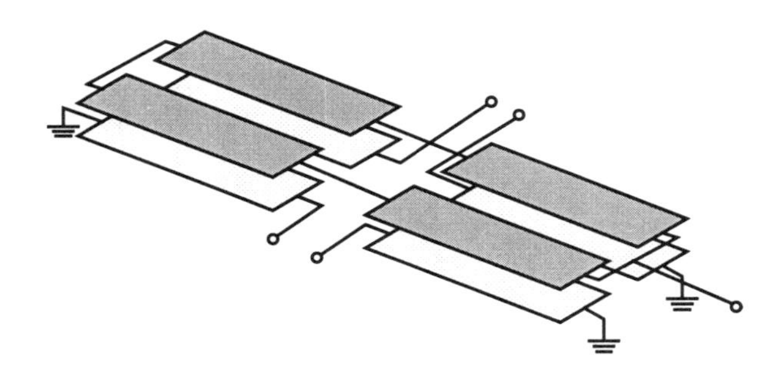

Figure 3.13 The three-conductor coupled-line RF or LO balun of the star mixer can be modeled as a set of simple coupled lines in parallel. This is not equivalent to the star-mixer balun, but it usually is an adequate approximation.

IF Design

The IF of this mixer requires no special design effort other than the obvious need to minimize the length of the jumper connection to the diodes. Connect it and forget about it!

3.6.4 Variations

Monolithic versions of this mixer work quite well. Simply move the two coupled lines from the underside of the substrate to the top side. The model in Figure 3.13 is still applicable; the lines in the monolithic circuit are edge coupled, not broadside coupled, as in the figure. Figure 3.14 shows one successful monolithic realization.

3.6.5 Cautions

This mixer is inordinately sensitive to the ground connection between the housing and the ground surface of the suspended substrate. Make sure that the substrate ground plane is well soldered to the housing around its entire circumference. If it is not, spurious resonances might appear in the RF/LO passband.

The housing for this mixer is necessarily about one-half wavelength square. This results in a wonderful environment for housing resonances. It may be necessary to mount some absorbing material in the housing to eliminate such resonances.

As with the other suspended-substrate mixers, do not use a high-dielectric-constant substrate. Doing so results in a mismatch between even- and odd-mode phase velocities in the coupled lines. As in all baluns, the phase-velocity inequality degrades the port-to-port isolation.

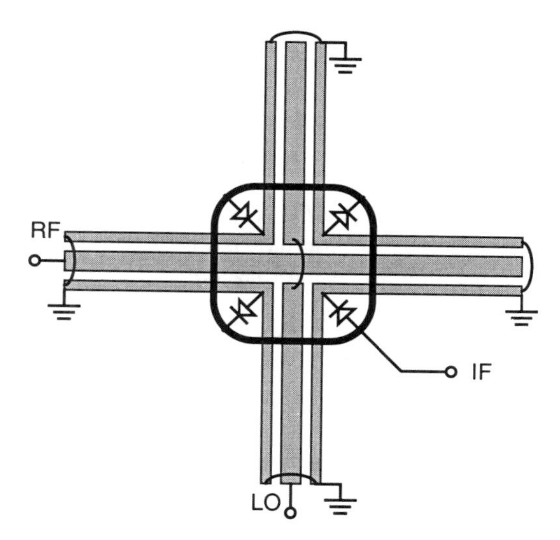

Figure 3.14 A practical monolithic realization of the star mixer. The conductors from the bottom side of the substrate in Figure 3.11 are moved to the top side of the substrate and are edge-coupled to the top-side conductors. Because of the increased complexity in the center of the structure, the diodes are connected to a ring instead of a common point. The IF is connected to that ring.

3.7 MONOLITHIC CIRCUITS

The classical diode-mixer circuits described in this chapter are difficult to realize as monolithic circuits. The fundamental difficulty is in realizing a high even-mode impedance in coupled-line baluns on a thin, high-dielectric-constant semiconductor substrate. This problem has two practical solutions: (1) use a structure that is more tolerant of low even-mode impedance, and (2) increase the even-mode impedance by increasing the even-mode inductance instead of reducing the even-mode capacitance.

The Marchand balun (see Figure 3.12 (a)) is far more tolerant of low even-mode impedance than the parallel-coupled balun used in the ring mixer (see Figure 3.6). Marchand baluns work well if the ratio of even- to odd-mode impedance is above approximately 3:1. The parallel-coupled balun requires at least a 10:1 ratio. If the frequency is low enough and the parallel-coupled balun is long enough, it can be wrapped in a spiral to increase its even-mode inductance. This has surprisingly little effect on the odd-mode impedance.

Coupled lines in monolithic mixers must be edge-coupled. Some types of mixers require very low odd-mode characteristic impedances, which are easier to realize as broadside-coupled suspended-substrate lines than in edge-coupled form. To achieve a low odd-mode impedance in a monolithic mixer, multistrip coupled lines can be used.

REFERENCES

[1] Maas, S. A., *Microwave Mixers*, 2nd ed., Norwood, MA: Artech House, 1993.

[2] Kollberg, E., *Microwave and Millimeter-Wave Mixers*, New York: IEEE Press, 1984.

[3] Kollberg, E., "Mixers and Detectors," in K. Chang, ed., *Handbook of Microwave and Optical Components*, New York: Wiley, 1990.

[4] Matthaei, G., L. Young, and E. Jones, *Microwave Filters, Impedance-Matching Networks, and Coupling Structures*, Norwood MA: Artech House, 1980.

[5] March, S., "A Wideband Stripline Hybrid Ring," *IEEE Trans. Microwave Theory Tech.*, Vol. MTT-16, June 1968, p. 361.

[6] Wadell, B. C., *Transmission Line Design Handbook*, Artech House, Norwood, MA, 1991.

Chapter 4
Diode Frequency Multipliers

A long time ago, when microwave components were made in a machine shop instead of a GaAs foundry and were designed with slide rules and Smith charts instead of $40,000 computer programs, designing a diode frequency multiplier was a true test of manhood.[1] These components genuinely deserved their reputation as troublemakers, and getting one to work required more wizardry, persuasion, and tolerance for frustration than scientific knowledge. The "multiplier guy" was the high priest of the department, and he deserved to be.

We might be justified in expecting that the occult art of multiplier design would be banished to the lore of the past by now. It hasn't. Diode frequency multipliers are still with us because they still do certain tasks better than active components: high-order multiplication, low noise, and broad bandwidths. Unfortunately, they're still just as difficult to produce. But here's how to do it.

Table 4.1 lists the circuits described in this chapter. We emphasize the resistive multipliers; because of their broad bandwidths, these probably are the most practical in modern microwave and RF systems. Varactor multipliers still have their niche, of course; their greatest benefit is low noise and operation at millimeter wavelengths.

4.1 FREQUENCY-MULTIPLIER THEORY

There are a number of efficient ways to generate harmonics with passive devices. (We'll leave the subject of active multipliers to Chapter 8.) The dominant devices are resistive diodes—Schottky-barrier diodes—and nonlinear-capacitance diodes, usually called *varactors* and *step-recovery diodes*. Let's examine our options.

1. In all fairness, it may have been a true test of womanhood, too, but back then there weren't many women working in the microwave industry. So we'll never know.

Table 4.1 Multipliers Described in This Chapter

Multiplier Type	Characteristics	Typical Applications
Single-diode resistive doubler	Simple, narrowband; approximately 10-dB conversion loss.	Low-cost, low-performance, "cheap and dirty" applications.
Singly balanced doubler with an output balun	Simple circuit; bandwidth depends on the balun. The balun must have a low impedance at the fundamental frequency, or effective bypassing is required.	General applications where the high conversion loss is tolerable.
Singly balanced doubler with an input balun	Simple circuit; bandwidth depends on the balun. The balun must have a low impedance at the harmonic. Since this is difficult to guarantee, bypassing invariably is necessary.	General applications where the high conversion loss is tolerable.
Doubly balanced resistive doubler	Requires input and output baluns, but no bypassing is needed, regardless of balun characteristics. Can be very broadband. Conversion loss of 10–13 dB, but higher power than the singly balanced circuits.	General applications where the high conversion loss is tolerable. With ~16 dBm input, output level may be adequate for a mixer LO.
Antiparallel-diode tripler	Broadband, low efficiency. Conversion loss is 13–16 dB. Fundamental-frequency output is not inherently rejected.	Applications where *really* high conversion loss is tolerable.
Varactor frequency doubler	Narrowband; sensitive to parameter variations; requires dc bias; can be unstable. Conversion loss of 3–5 dB, depending on frequency. Not for the faint of heart.	Mixer LO, especially at millimeter wavelengths. Used primarily at frequencies where active or resistive multipliers are impractical.
Step-recovery diode multiplier	High-order frequency multiplication from a relatively low-frequency source.	Mixer LO systems, frequency synthesizers, other signal sources.

4.1.1 Resistive Frequency Multipliers

Resistive frequency multipliers use the nonlinear I/V characteristic of a Schottky-barrier diode to distort a sinusoidal waveform. This distortion generates harmonics. That's about all there is to it. Designing a resistive multiplier simply reduces to selecting optimum source and load impedances and separating the input and output signals. This is usually done most effectively by a balanced circuit.

General Characteristics

Clearly, the more you distort the input sinusoid, the greater the harmonic currents in the diode. When the diode rectifies the sinusoid, much as a diode in a dc power supply, the distortion is about as great as it can be, and the harmonic output is maximized. Unfortunately, that maximum still is not very great. Resistive frequency multipliers are not very efficient.

How good can the efficiency be? More to the point, how bad does it have to be? Page [1] proved that the optimum efficiency of a resistive frequency multiplier can be no greater than $1/n^2$, where n is the harmonic number. A doubler, for example, must have at least 6-dB conversion loss; a tripler, almost 10 dB. In reality the story is even worse: doublers rarely have conversion loss below about 10 dB, and no one bothers to make higher-harmonic resistive multipliers. It's just too depressing!

Resistive multipliers, however, have one significant advantage over reactive multipliers: broad bandwidth. Since a Schottky diode is a resistive device, it is inherently very broadband. As with diode mixers, the bandwidth of these components is limited primarily by the external circuitry, usually baluns or hybrids, and not by the diodes themselves. Multi-octave resistive multipliers are regularly produced.

Design Theory

A simple, semi-empirical analysis of a single-device resistive frequency doubler is given in [2]. As with the balanced mixers examined in Chapter 3, balanced multipliers are essentially just interconnections of single-device multipliers, so the approach is valid for balanced circuits as well.

The basic multiplier is shown in Figure 4.1. We assume that (1) the diode's series resistance is small, and (2) the diode's capacitance can be "absorbed" into the capacitance of the resonators. These resonators, marked ω_1 and $2\omega_1$, are ideal parallel-resonant circuits, so they provide open circuits at resonance and short circuits at all other frequencies.[2] This is the optimum termination.

The analysis in [2] gives the following results. The junction resistance, R_j, is

2. This appears to conflict with our grand assertion that resistive multipliers are broadband. Practical multipliers invariably are balanced, and their terminations are provided automatically by the topology of the circuit. We don't need to use resonators, or even filters.

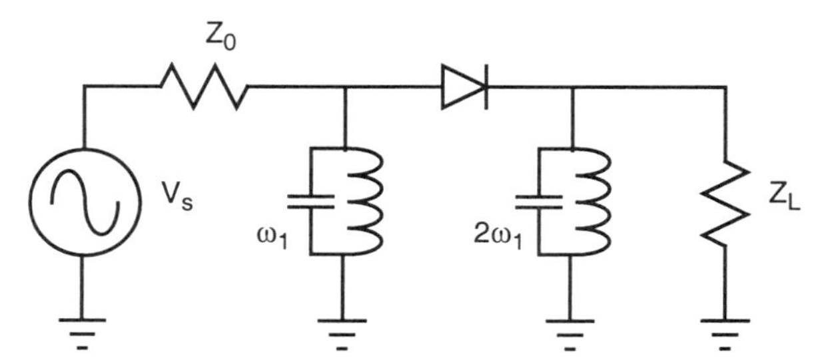

Figure 4.1 Simplified model of a resistive diode frequency doubler. The blocks marked ω_1 and $2\omega_1$ are ideal parallel resonators at the input and output frequencies, respectively.

defined as the ratio of the fundamental-frequency junction voltage and current and is given by

$$R_j = \frac{V_s}{I_{max}} - R_s \tag{4.1}$$

where V_s is the magnitude of the source voltage and I_{max} is the peak junction current. V_s is determined by the available power, and I_{max} is determined by the limitations of the diode, so R_j is established at the outset. V_s is

$$V_s = \sqrt{8Z_0P_{av}} \tag{4.2}$$

where P_{av} is the available power from the source and Z_0, the source impedance, is assumed to be real. The input impedance becomes

$$Z_{\text{in}} = R_j + R_s \tag{4.3}$$

where R_s is the diode's series resistance. When the input is conjugate matched, $Z_0 = Z_{\text{in}}$, P_{av} equals the input power, P_{in}, and

$$P_{av} = P_{in} = \frac{1}{8}I_{max}^2(R_j + R_s) \tag{4.4}$$

The optimum load impedance, Z_L, is

$$Z_L = 0.83(R_j - 2R_s) \tag{4.5}$$

and the output power, P_L, is

$$P_L = 0.0167I_{max}^2(R_j - 2R_s) \tag{4.6}$$

That very small factor of 0.0167 should be a little disturbing. Dividing (4.6) by (4.4), we obtain the conversion loss,

$$G_c = 0.13\left(\frac{R_j - 2R_s}{R_j + R_s}\right) \tag{4.7}$$

which predicts a loss of at least 9 dB for the practical multiplier. Ouch!

This is a little disturbing. After all, Page's theorem says that we are limited to 6-dB loss, and it seems reasonable to be able to come closer to this limit. Part of the problem is the assumption, used to derive (4.7), that the diode-current waveform is a rectified sinusoid. A more rectangular waveform would have better efficiency, and nonlinear analyses on the computer confirm this. Even with the strong diode nonlinearity, however, such a waveform is difficult to achieve. On the other hand, some limitations are not included in (4.7), the most important being current saturation. Current saturation limits the peak current and increases the series resistance at high current densities, reducing efficiency and output power.

The design process follows this analysis. We first determine V_s and I_{max} from the available input power and the maximum current the diode can handle. V_s may also be limited by the breakdown voltage of the diode. We then find R_j from (4.1), and we obtain the input impedance, optimum load impedance, and conversion loss or output power from (4.3) through (4.7). Finally, we design the rest of the circuit—baluns or filters, depending on the type of multiplier desired.

4.1.2 Varactor Multipliers

A nonlinear reactance also can distort a sinusoidal signal. The nonlinear reactance most readily available is the depletion capacitance of a Schottky or *pn* junction diode. This nonlinearity is not very strong, so it is difficult to obtain good efficiency. Nevertheless, with a degree of care comparable to balancing a needle on its point, an efficient multiplier can be produced.

General Characteristics

The advantages and disadvantages of a varactor multiplier are the opposite of those of the resistive multiplier. A varactor is capable of higher efficiency and power than a resistive multiplier, theoretically 100% for all harmonics. However, varactor multipliers are notoriously narrowband. This is to be expected, since the varactor's susceptance must be great enough, at the input frequency, to allow substantial reactive current.

One troublesome characteristic of varactor multipliers is their extreme sensitivity to almost every parameter of the circuit. These components have high design sensitivity: very small changes in circuit parameters (tuning reactances, bias voltage, input power level, and similar quantities) change the output power substantially.

Making a varactor multiplier work—and keeping it working—can be a delicate business; it's not for the easily frustrated. Varactor frequency multipliers require a lot of empirical tuning; you can't just design one, build it, and expect it to work. For this reason, it is rare to see a monolithic varactor multiplier. The uncertainties in monolithic circuit design just don't allow it.

Idlers

One problem in designing a varactor multiplier is the diode's weak reactive nonlinearity. Without some tricks, the multiplier does not generate harmonics efficiently beyond the second. To see why, consider the varactor's capacitance-voltage (C/V) characteristic:

$$C_j(V_j) = \frac{C_{j0}}{\sqrt{1 - \frac{V_j}{\phi}}} \tag{4.8}$$

where the quantities are as described in Section 2.1.2. The charge-voltage characteristic is found by integrating (4.8):

$$Q(V_j) = -2C_{j0}\phi(1 - V_j/\phi) \tag{4.9}$$

This can be rearranged to express the voltage as a function of current:

$$V_j = \phi\left(\frac{Q_\phi^2 - Q^2}{Q_\phi^2}\right) \tag{4.10}$$

where Q is the junction charge and $Q_\phi = -2C_{j0}\,\phi$, a constant. If the current in the diode is sinusoidal, the charge function will be also, and the junction voltage will be roughly proportional to the square of the charge. This means that the varactor multiplier will generate only second harmonics efficiently. For higher harmonics, we need to do more. The standard trick is to add an *idler*, a short-circuit resonator tuned to the second harmonic, in parallel with the diode. This resonator allows a second-harmonic current to circulate through the diode and to mix with the fundamental, producing a third harmonic. If yet higher harmonics are required, idlers at the other intermediate harmonics can be included. However, for high harmonics, it usually is better to cascade two or more low-harmonic multipliers. The efficiency is better than in a single multiplier, and it's much easier to cascade stages than to implement a large number of idlers. In any case, the basic circuit of a single-diode multiplier is shown in Figure 4.2.

Idlers can be implemented in a number of ways. One method is to exploit a resonance in the varactor's package; a little extra capacitance can be added, if

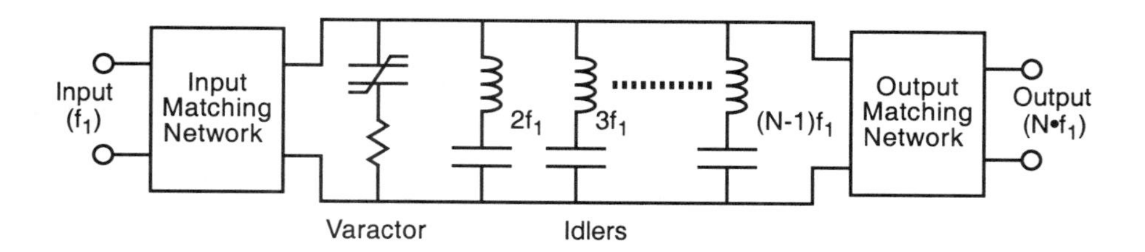

Figure 4.2 Basic circuit of a varactor frequency multiplier, showing the matching circuits, varactor, and idlers. In high-order multipliers, not all intermediate frequencies need have idlers, but, theoretically, efficiency will be highest if they do.

necessary, to adjust the resonant frequency. Other methods are the obvious: a stub or waveguide resonator. It just depends on the type of structure selected for the multiplier.

Diodes

Multipliers use either *pn*-junction or Schottky varactors, either GaAs or silicon. Their characteristics are as follows:

- *pn-Junction varactors:* Most multipliers use *pn* junction varactors. Strangely, the characteristic that makes a *pn* junction unsuitable for mixers—diffusion charge storage in forward conduction—makes it very useful for multipliers. In a reactive multiplier, we want charge to be stored in the junction, not conducted; conduction causes power dissipation, hence loss and heating. Charge storage, instead of conduction, prevents this and actually increases the capacitive nonlinearity. When a diode is used in this manner, it is said to be *overdriven*.
- *Punch-through varactors: pn*-junction varactors sometimes are designed to be fully depleted at modest reverse voltages. When such *punch-through* varactors are used, the multiplier's output power does not vary much with input power. This clearly has practical value, but it also has a cost: slightly lower efficiency.
- *Schottky-barrier varactors*: Diffused *pn*-junction varactors are limited in capacitance to about 0.2–0.3 pF or more. This is too great for millimeter-wave multipliers. The only real option is to use a Schottky diode and to forego the benefits of diffusion charge storage. The disadvantage of a Schottky varactor is that the user must be careful not to overdrive the multiplier; if the multiplier is driven to the point where rectification occurs, the efficiency and output power drop substantially.
- *GaAs vs. silicon*: The usual trade-off: GaAs is more expensive but offers better performance It is usually reserved for situations where it is really necessary: high frequencies and high efficiencies.

Burckhardt Theory

Much varactor frequency-multiplier design is based on a classic paper by C. B. Burckhardt [3]. Burkhardt derived tables of design data for optimum frequency multipliers having various orders and combinations of idlers.

In Burckhardt's scheme, the input and output ports of the multiplier are modeled as shown in Figure 4.3. S_{01} and S_{0N} are the input and output elastance, or inverse capacitance; N is the harmonic number; while R_i and R_L are the real parts of the input and optimum load impedance, respectively. Knowing these, we can design the matching circuits. Of course, we also must determine the input and output power levels. These are functions of the normalized drive level, D, defined as

$$D = \frac{q_{max} - Q_B}{q_\phi - Q_B} \tag{4.11}$$

where q_{max} is the maximum junction charge, including diffusion charge; Q_B is the charge at breakdown voltage, V_b, which we assume to be the peak reverse drive voltage; and q_ϕ is the charge when the junction voltage equals the built-in voltage, ϕ. If $q_{max} = q_\phi$, the drive level is as great as it can be without causing diffusion charge storage, and $D = 1.0$. If $D > 1.0$ there is some degree of diffusion charge storage, and the varactor is overdriven.

Tables 4.2 and 4.3 present design data from [3]. These cases, a doubler and tripler using abrupt-junction diodes, are probably the most useful. In these tables $S_{max} = 1 / C_{min}$, where C_{min} is the varactor capacitance at V_b. α and β are used to determine the conversion efficiency and output power. The efficiency is

$$G_c = \exp\left(\frac{-\alpha}{Q_\delta}\right) \tag{4.12}$$

where

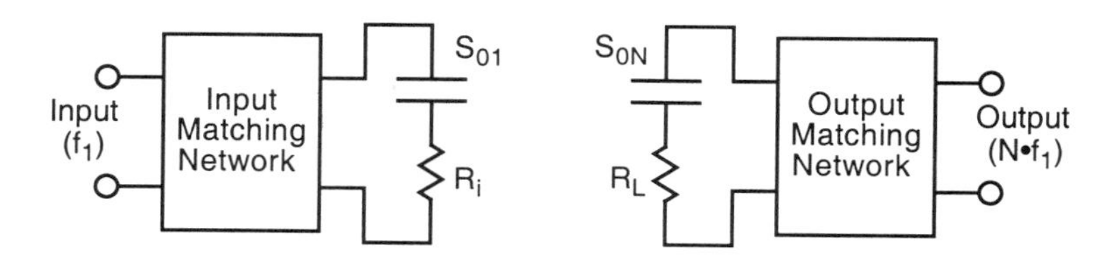

Figure 4.3 Input and output models used by Burckhardt for design of the multiplier's matching networks.

Table 4.2 Doubler

Parameter	D = 1.0	D = 1.3	D = 1.6
α	9.95	8.3	8.3
β	0.0227	0.0556	0.0835
$R_i\omega_1/S_{max}$	0.080	0.098	0.0977
$R_L\omega_1/S_{max}$	0.1355	0.151	0.151
S_{01}/S_{max}	0.50	0.37	0.28
S_{02}/S_{max}	0.50	0.40	0.34
$V_{dc,n}$	0.35	0.28	0.24

Table 4.3 Tripler

Parameter	D = 1.0	D = 1.3	D = 1.3
α	11.6	9.4	9.8
β	0.0241	0.0475	0.070
$R_i\omega_1/S_{max}$	0.137	0.168	0.172
$R_L\omega_1/S_{max}$	0.0613	0.0728	0.0722
S_{01}/S_{max}	0.50	0.36	0.26
S_{02}/S_{max}	0.50	0.38	0.31
S_{03}/S_{max}	0.50	0.38	0.30
$V_{dc,n}$	0.32	0.24	0.18

Source: C. B. Burckhardt, "Analysis of Varactor Frequency Multipliers for Arbitrary Capacitance Variation and Drive Level," *Bell System Technical Journal*, Vol. 44 (April, 1965), p. 675. © 1965 AT&T. Reprinted from the *Bell System Technical Journal* with permission.

$$Q_\delta = \frac{S_{max}}{\omega_1 R_s} \tag{4.13}$$

R_s is the diode's series resistance, and ω_1 is the input radian frequency. The output power is

$$P_L = \beta\omega_1\left(\frac{\phi - V_b}{S_{max}}\right) \tag{4.14}$$

S_{02}, given in the second table, is the component of diode capacitance at the second harmonic. It is used for designing the idler.

The tables also give values for the normalized bias voltage, defined as

$$V_{dc,\,n} = \frac{\phi - V_{dc}}{\phi - V_b} \tag{4.15}$$

where V_{dc} is the dc bias voltage. The dc bias invariably is treated as an empirical tuning element for optimizing the multiplier, so this quantity is not especially useful.

4.1.3 Step-Recovery-Diode Multipliers

Wouldn't it be nice if there were some way to get around the limitation of the very weak depletion-capacitance nonlinearity? One possibility is to exploit the diffusion charge storage of a *pn* junction when the diode is forward biased. This is indeed a strong nonlinearity; the stored charge is

$$Q_{diff} = \tau I_d \approx \tau I_s \exp(\delta V) \tag{4.16}$$

which is an exponential nonlinearity, much stronger than that wimpy little depletion capacitance in (4.9). Now, if we drive the diode through an appropriate inductor, we can arrange things so the diode is completely discharged at the instant when the inductor current is maximum. At that point, we have an inductor with a huge current, thanks to the large stored diffusion charge, driving a small depletion capacitance. We then get precisely one-half cycle of voltage—a fast pulse—before the diode becomes forward biased again, and the process is repeated. Applying this pulse to a resonator is like hitting a church bell with a hammer: it will ring at the resonant frequency. Of course, that resonator should be tuned to a harmonic of the pulse's fundamental frequency. *Voila*: a high-order frequency multiplier.

SRD Multiplier Basics

The standard analysis of the SRD multiplier is in [4,5]. The diode in those analyses is idealized: it is a short circuit when the current is forward, virtually all forward current is stored, and the diode has a constant capacitance when reverse biased.

Figure 4.4(a) shows the circuit of the SRD pulse generator and Figure 4.5 shows the waveforms in the circuit. The impedance Z_s in Figure 4.4(a) has a finite value at the fundamental frequency and is a short circuit at all other harmonics. In practice, some sort of filter is necessary. When the current in the diode is in the forward direction, the diode voltage is (ideally) zero, and charge is stored in the diode's junction. The diode current is equal to the inductor current, $I_L(t)$; the equivalent circuit is shown in Figure 4.4(b). When the current reverses, the diode voltage remains at zero until all the charge is removed; then the diode switches abruptly to a low-capacitance state, the minimum depletion capacitance. Now, this capacitance is driven by the peak current in the inductor, and since the diode was able to store a lot of charge, this current is quite high. The equivalent circuit, shown in Figure 4.4(c), is simply a resonator driven by the inductor current. The resulting damped oscillation lasts only one-half cycle; however, before the diode voltage, V_d, again becomes positive, the diode turns on, and the oscillation is quenched. The result is a large, very fast pulse, which can be as short as a few tens of picoseconds.

This fast pulse is very useful. It can be applied to a resonator and a high harmonic obtained. It also can be used for a variety of switching and sampling

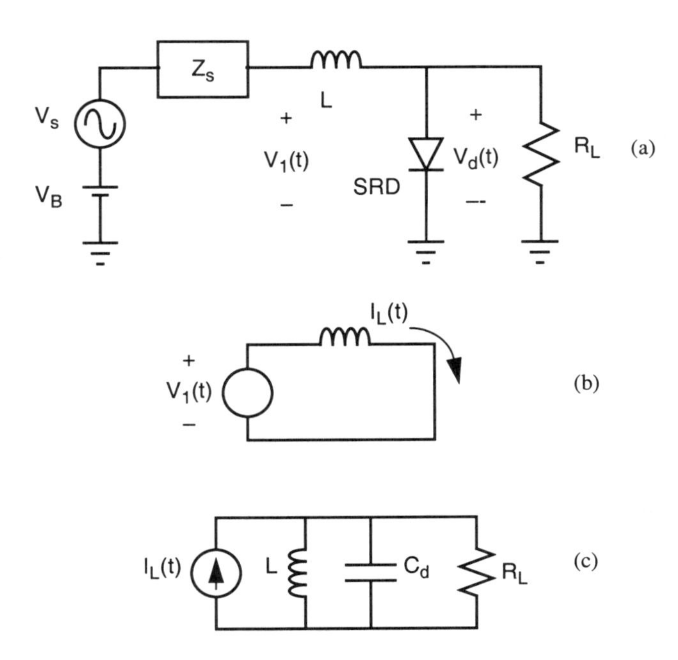

Figure 4.4 (a) Step-recovery diode multiplier circuit; (b) equivalent circuit when the diode is forward biased; and (c) equivalent circuit during the impulse interval.

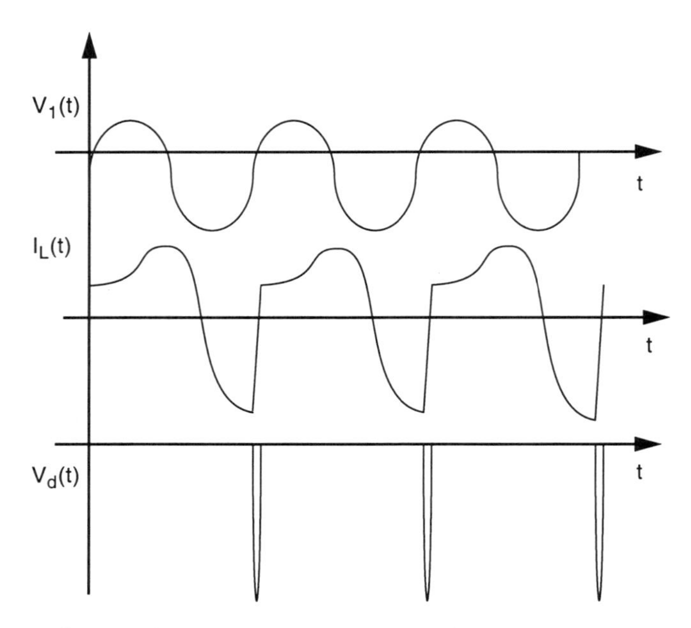

Figure 4.5 Waveforms in the step-recovery diode circuit shown in Figure 4.4.

applications. Finally, part or all of its spectrum can be used as a "comb" of frequencies for various purposes.

The SRD multiplier is fundamentally a reactive multiplier. Therefore, it is not subject to the $1/n^2$ efficiency limitation suffered by resistive multipliers. This is not to say that high-harmonic SRD multipliers are efficient, however; in practice, their efficiency is not much better than $1/n^2$, but unlike resistive multipliers, at least it's not much worse.

Design Theory

As with other such circuits, we need to know (1) input impedance, (2) optimum load impedance, (3) input and output power levels, and (4) how to select circuit and diode parameters to achieve these. From Figure 4.4, we see that the design involves primarily selecting the diode, inductance, and load resistance. The pulse is a half cycle of a sinusoid, and we want it to be approximately one-half the period of the output sinusoid. Its length is

$$T_p = \frac{\pi}{\omega_n} \tag{4.17}$$

where ω_n is the natural frequency of the damped sinusoid,

$$\omega_n = \sqrt{\frac{1-\varsigma^2}{LC_d}} \tag{4.18}$$

ζ is the damping factor,

$$\varsigma = \frac{1}{2R_L}\sqrt{\frac{L}{C_d}} \tag{4.19}$$

ζ must be selected carefully. If ζ is too low, the multiplier can become unstable; if it is too high, the pulse will be unnecessarily long. To guarantee stability, ζ should be no lower than 0.4–0.5. These considerations largely define the values of L, C_d, and R_L.

To a good approximation, the imaginary part of the multiplier's input impedance is simply $j\omega_1 L$, the inductor's reactance at the fundamental frequency, ω_1. The real part is somewhat more difficult to determine. A useful expression is

$$R_{in} = \frac{4V_p^2 L^2 \omega_1^3}{V_1^2 R_L \omega_n} \tag{4.20}$$

where V_p is the peak pulse voltage,

$$V_p = -I_0\sqrt{\frac{L}{C_d}}\exp\left(\frac{-\pi\varsigma}{2\sqrt{1-\varsigma^2}}\right) \tag{4.21}$$

and I_0 is the peak value of $I_L(t)$. The input power is

$$P_{in} = \frac{V_p^2\omega_1}{4R_L\omega_n} \tag{4.22}$$

The output power is difficult to estimate with any degree of accuracy. Certainly, we could determine the power in the Nth harmonic of a sinusoidal pulse train, and estimate the output power as that quantity. That would be valid if all the higher harmonics were terminated in the resistance R_L. In real multipliers, however, many harmonics are reactively terminated, and this allows some of their power to be "recycled" in the nonlinear diode and converted to output power at the desired harmonic. Furthermore, other phenomena, such as losses and limitations in the switching time of the diode, strongly affect high harmonics. Output power is difficult to predict accurately. For better or worse, the best way to determine the output power is to build the circuit.

4.2 SINGLE-DIODE RESISTIVE FREQUENCY DOUBLER

4.2.1 Characteristics

The single-diode multiplier is probably not the preferred way to make one of these circuits. In a single-diode circuit we depend heavily on filters to provide fundamental-frequency rejection; in balanced circuits, the fundamental is rejected automatically. Still, as with single-diode mixers, single-diode multipliers have their uses and serve as a prototype for more complex circuits.

As we noted in Section 4.1.1, resistive multipliers represent one extreme of the efficiency/bandwidth trade-off; varactor multipliers represent the other. The middle ground is occupied by active multipliers. Resistive multipliers are inefficient, a doubler rarely having conversion loss better than 10 dB.

The single-diode multiplier is useful mainly for low-cost, low-performance applications, where the cost of two or four diodes for a balanced multiplier might be prohibitive. Another application is at high frequencies, in waveguide structures, where waveguide bandwidths are possible and the fundamental frequency is easy to reject (by keeping it below the waveguide's cutoff frequency). A single-diode circuit does have one advantage over multiple-diode circuits: it is easy to provide dc bias to it. Dc bias allows an extra degree of freedom in optimizing the multiplier. It is much more difficult (but not impossible) to bias a balanced multiplier.

4.2.2 Description

Figure 4.6 shows the basic circuit. It consists of a diode and two resonators tuned to the fundamental and second harmonic. The figure shows parallel LC resonators, but any type of filter having appropriate characteristics can be used.

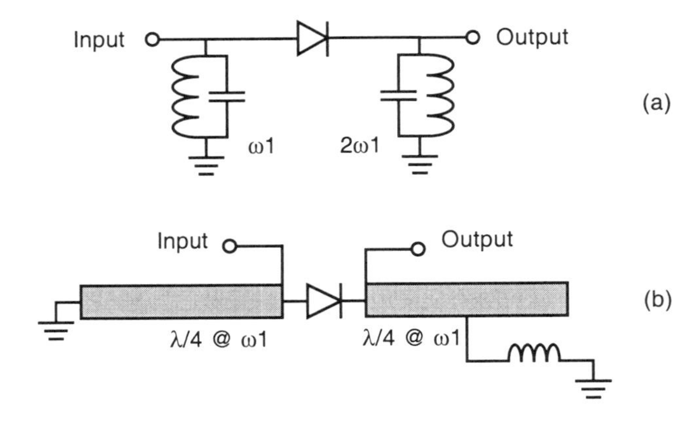

Figure 4.6 Single-diode resistive frequency doubler: (a) lumped realization; (b) distributed. The inductor in the distributed circuit is a dc current return.

The parallel LC resonators are ideal because they short-circuit the diode at the unwanted harmonics, giving the best overall performance. Note that the resonators conveniently decouple the input from the output, and put the diode in parallel with the input at the fundamental frequency and in parallel with the output at the second harmonic. The inductors in the resonators can be tapped to optimize the source and load impedances at the diode.

4.2.3 Design

The design theory for this multiplier is in Section 4.1.1 and the relevant equations are (4.1) through (4.7). The first step is to determine a value of I_{max} and V_s and then the diode's effective junction resistance, R_j, from (4.1). I_{max} is limited to the diode's peak current capability and V_s to its reverse breakdown voltage. The input power, output power, input impedance, and optimum load impedance are then calculated from (4.3)–(4.7). If these are not what is desired, the process is repeated with different values of I_{max} and V_s.

The multiplier can be realized literally as shown in Figure 4.6 (a). The diode's junction capacitance is part of the resonator capacitance, and tap points in the inductor can be adjusted empirically to optimize the impedance presented to the diode and thus its performance.

At higher frequencies transmission-line stubs can be used. Both stubs should be one-quarter wavelength long at the fundamental frequency. This way, the short-circuit stub grounds the left end of the diode at the second harmonic, and the open-circuit stub grounds the right end at the fundamental frequency. Note that a dc-current return is needed. The best place to put it is a zero-voltage point at the second harmonic; we have such a point at the center of the output stub. For best fundamental-frequency rejection, the input and output ports should be connected as shown; some control of the impedances presented to the diode can be obtained by tapping these points lower on the resonator, but isolation will suffer. The lengths of the resonators may have to be adjusted somewhat to compensate for the diode's junction capacitance.

Computer analysis, either by harmonic-balance or transient methods, works well for resistive diode multipliers. It is especially helpful for optimizing the isolation, conversion efficiency, and port VSWR; these are competing trade-offs. We recommend it heartily.

Diode Selection

Diode selection for resistive multipliers is similar to that for mixers. The junction capacitance should be low enough to be negligible at the output frequency. At high frequencies, this may not be possible. In this case the designer must trade off diode capacitance against output power; larger diodes can handle higher power but have higher capacitance. See Section 4.2.4 for more ways to improve power handling.

Resistive multipliers are best fabricated from GaAs diodes. Silicon diodes usually have very low reverse-breakdown voltages, which limit their output power and may result in an impractically low input impedance. The higher series resistances of silicon diodes results in higher conversion loss, exacerbating an already dismal situation.

Remember that very little input power is converted to output power, so the diode must dissipate virtually all the input power. Conventional GaAs Schottky diodes rarely can dissipate more than 20 dBm safely. Higher power levels usually require a multiple-diode circuit.

4.2.4 Variations

At high frequencies, the designer faces a dilemma: it may be impossible to find a diode that has negligible junction capacitance. He can adjust the lengths of the stubs to compensate, but once this is done, the stubs no longer present the proper short- and open circuit-terminations to the diode at the first and second harmonics.

One way around this dilemma is to use two large diodes in series. Doubling the area of the diode junction doubles I_{max}, thus doubling the power, but also doubling the junction capacitance. Putting two of these diodes in series halves their capacitance, returning the original capacitance, but doubles the breakdown voltage. This again doubles the power, so the same conversion loss, input impedance, and load impedance are obtained at 6-dB increased input and output power.

Putting two diodes in series may not precisely double the breakdown voltage. Differences in the diodes' reverse leakages may cause the voltage to divide between the diodes unequally. The situation is not as bad as it might appear, because the impedance of the junction capacitance usually is much less than the impedance of the reverse-biased junction, so the voltage-divider effect of the capacitors tends to equalize the voltage across the two diodes. In any case, the breakdown voltage of the diode pair still will be greater than that of a single diode.

The configurations in Figure 4.6 are not the only ones possible, of course. Filters can be used in place of the resonators to achieve broader bandwidth. It is essential, however, to short-circuit the diode at unwanted harmonics, and conventional filter designs may not do this. It also is essential that the input filter not load the diode at the output frequency or the output filter at the input frequency. This is a difficult set of requirements to meet. Faced with so much trouble, you might find that a balanced multiplier makes more sense.

4.2.5 Cautions

All in all, this is not a very good circuit. Balanced multipliers almost always are preferred. You should use it only if you understand the advantages of balanced multipliers and are sure that a balanced circuit is clearly inappropriate for your application.

4.3 SINGLY BALANCED FREQUENCY DOUBLER: INPUT BALUN

4.3.1 Characteristics

One of the most important characteristics of a balanced multiplier is its inherent rejection of the fundamental-frequency output. This is especially important in resistive multipliers, because their efficiency is low. Without substantial rejection, the fundamental-frequency output power can be much greater than the harmonic.

The use of a balanced circuit unfortunately does not improve the conversion efficiency of the multiplier. By power-combining two or more diodes, however, it does improve the output power by 3 dB. Of course, 3 dB greater input power is needed.

Singly balanced circuits require a balun at either the input or the output; doubly balanced multipliers require baluns at both the input and the output. Avoiding one balun seems at first to simplify the circuit; however, it often results in a situation where there is no current return for the second harmonic. Providing a current return may complicate the circuit and reduce its bandwidth enough to make a doubly balanced mixer look very attractive.

4.3.2 Description

Figure 4.7 shows the prototype circuit of a singly balanced multiplier using an input transformer as a balun. It looks a lot like a dc power-supply circuit. In a power supply we are interested in the dc output, and we short-circuit the harmonics; in the doubler, we are interested in the second harmonic and short circuit the dc current.

In the prototype circuit, the transformer's center tap provides a current return for both the dc and even-harmonic currents; the inductor is an RF choke, preventing dc current from circulating in Z_L. The odd-harmonic currents simply circulate in the loop that comprises the diodes and transformer secondary. (Before you ask: how can the diode carry a reverse current? It doesn't. Remember, the rectifying property of the diode applies to the *total* junction current, not to individual harmonics. The total

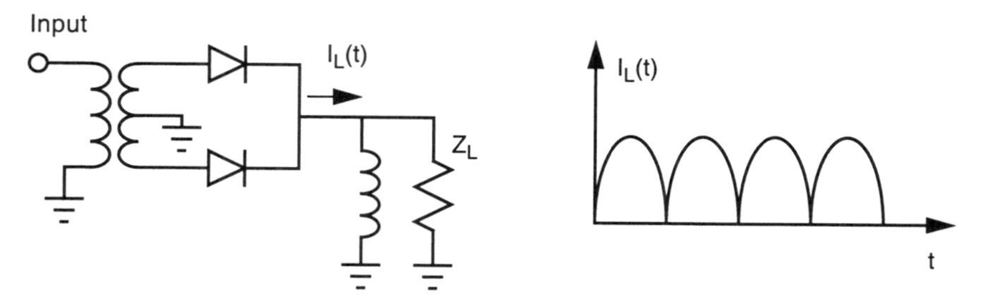

Figure 4.7 Prototype circuit of the doubly balanced resistive doubler and its output current. The current waveform has only a dc component and even harmonics.

current in each diode does indeed satisfy this requirement.) In a high-frequency circuit, a balun replaces the transformers; most types of baluns have no acceptable high-frequency current-return path. In this case we must provide one or find a different balun. This is a fundamental problem in designing singly balanced multipliers.

4.3.3 Design: Rat-Race Multiplier

Figure 4.8 shows a practical high-frequency balanced doubler. It uses a rat-race hybrid (discussed in Section 3.2.2) and simple stubs, each one-quarter wavelength long at the fundamental frequency, for current returns. The circuit looks very much like the rat-race mixer described in Section 3.2. Note, however, that the diodes are reversed, and there are a few other important differences.

The hybrid is good for about a 10% bandwidth, perhaps 15% if the port-VSWR requirements are modest. Because the diodes rarely limit the circuit's bandwidth, the bandwidth of the multiplier essentially is the bandwidth of the hybrid.

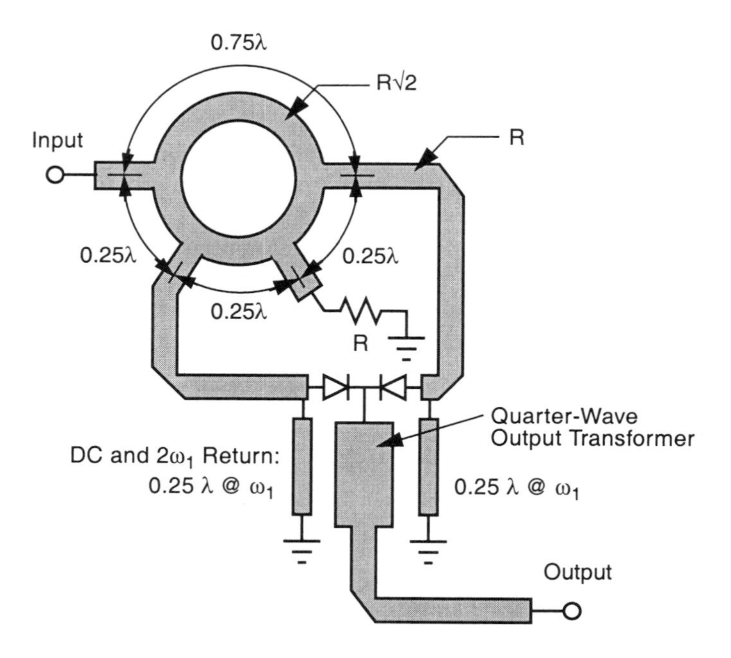

Figure 4.8 Balanced resistive doubler using a rat-race hybrid. The lengths of the transmission lines from the ring to the diodes must be equal.

Hybrid Design

Figure 4.8 largely defines the design of the hybrid. The impedance of the ring is 1.414R, where R is the port impedance, usually 50Ω. The impedance of the ports connected to the diodes is likewise R, so the individual multiplier diodes should be designed to have an input impedance of R. The lines from these ports to the diodes must be equal in length, and the remaining port must be terminated as shown.

Multiplier Design

The multiplier largely consists of two single-diode multipliers connected to a hybrid. (The design of single-diode multipliers is covered in Section 4.2.3.) The main difference between this circuit and the single-diode multiplier is that the source resistance for each diode is R, but, because the diodes are in parallel at the output frequency, the load resistance seen by each diode is $2R$. If R is 50Ω, the load resistance for each diode is 100Ω. Unfortunately, (4.5) shows that the load resistance should be roughly equal to the source impedance. Thus, it appears that we're stuck with a mismatch; our only option is to decide at which port we want to suffer it.

This is less of a problem than it appears. One solution is to match the input and to suffer some additional loss at the output, if necessary. A second solution is to use the lines from the balun to the diodes as quarter-wave transformers, and design the individual multipliers for a 100Ω input and output impedance. The change in source impedance may require reduced I_{max}, however, which decreases the output power, and the higher values of R_j may make junction capacitance a problem. A third option, perhaps the best, is to design for a 50Ω input impedance and to use a quarter-wave output transformer to transform the 50Ω load impedance to 25Ω at the diodes, giving each diode a 50Ω load. The latter approach is developed here.

In summary, here is the process:

1. Select a diode according to the criteria in Section 4.2.3.
2. Select values of V_s and I_{max} so that $R_j + R_s$ is approximately 50Ω; see (4.1).
3. Calculate the input power and output power from (4.4) and (4.6).
4. Design the hybrid and current-return stubs. The stubs' characteristic impedances should be approximately 50–70Ω.
5. Design the output quarter-wave transformer. The characteristic impedance should be approximately 35Ω.
6. Optimize the circuit with nonlinear analysis. Harmonic-balance analysis is the preferred method. If this is not available, at least use linear analysis to optimize the design of the hybrids. Use a 50Ω resistor in parallel with the diode's junction capacitance (C_{j0}) to model each of the diodes.

4.3.4 Design: Coplanar Multiplier

A clever circuit that solves the current-return problem was proposed by Ogawa and his colleagues [6]. It uses a coplanar-waveguide (CPW) structure and exploits the fact that the output currents at the fundamental frequency and second harmonic have different modes.

Figure 4.9 shows the circuit. The input is CPW, and it immediately transitions to a slotline. The slotline impedance should be the same as the CPW, usually 50Ω. To avoid coupling spurious modes near the junctions, and their resulting strange passband irregularities, the slotline should be at least one-eighth wavelength long. The diodes' cathodes are connected to a quarter-wave transformer, which can be used to optimize the load impedance.

The bond wires are essential. Wire 1 is a necessary part of the CPW-to-slotline transition; it provides a path for the currents in the left side of the CPW. The reason for wire 2 is a little more subtle. Although the diodes excite a normal CPW mode at the second harmonic, at the fundamental frequency they excite a mode propagating on the CPW line as if it were two slotlines side by side. Wire 2 short-circuits this slotline mode but not the normal CPW mode.

Wire 1 should be as close to the slotline junction as possible. Wire 2 should be located one-quarter of the fundamental-frequency wavelength from the diodes.

Remember, the diodes in this circuit are in series with the slotline at the input and in parallel with the CPW at the output. The source and load impedances must be adjusted for this.

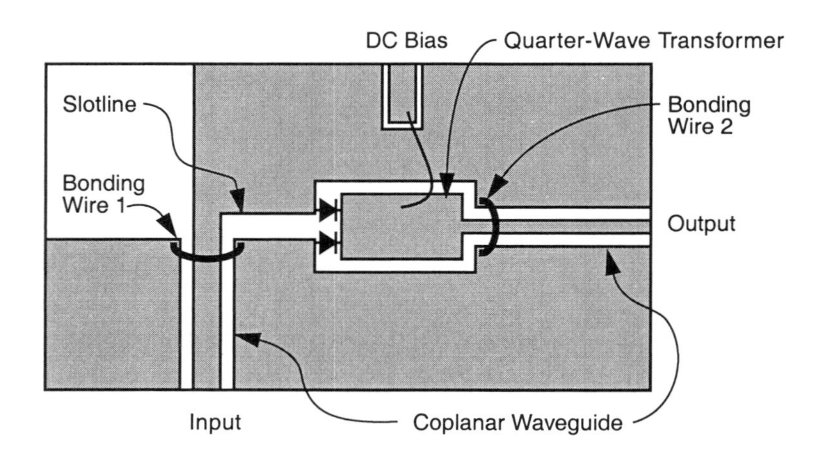

Figure 4.9 Balanced multiplier of Ogawa et al. [6]. The wide part of the output strip is a quarter-wave transformer for the second harmonic. The bonding wires are essential. Bonding wire 1 is part of the CPW-to-slotline transition, and wire 2 prevents fundamental-frequency leakage in the odd CPW mode. (Figure reprinted with permission, © 1987, IEEE.)

The process is as follows:

1. Create a single-diode prototype using (4.1) through (4.7) and the information in Section 4.2.3. The input impedance $R_j + R_s$ in (4.1) should be around 50Ω.
2. Design the slotline to act as a quarter-wave transformer between the 50Ω source and the input impedance of the diodes, 2 $(R_j + R_s)$.
3. Similarly, design the CPW output transformer to transform the 50Ω load impedance to $Z_L / 2$ at the diodes; Z_L is found from (4.5).
4. Locate the bond wire one-quarter wavelength from the diodes. It can be adjusted slightly to tune the multiplier.
5. Don't forget the bond wire to the dc bias pad. If dc bias is not used, ground this wire. It is the diodes' dc current return.

As in any CPW circuit, the ground currents should be in the CPW ground plane, not the microstrip ground plane under the substrate. Otherwise, microstrip modes will exist and the behavior of the circuit will be strange indeed. For this reason a 2.5-mm thick, low-dielectric-constant substrate having a dielectric constant below 3.0 is ideal; an air gap between the substrate and the ground plane, to increase the impedance of microstrip modes, is even better.

4.3.5 Variations

The most useful variation is the singly balanced doubler with an output balun. This circuit is described in Section 4.4.

Many other types of baluns can be used instead of a rat-race hybrid. For broadband multipliers, a Marchand balun works well and provides a much wider bandwidth; for design information, see Section 3.6.3. Don't count on the Marchand balun to provide a second-harmonic current return; you still need a set of stubs.

4.3.6 Cautions

In the circuit in Figure 4.8, it is tempting to use the current-return stubs to resonate the diodes' junction capacitances. Doing so requires reducing the length of the stubs, and they no longer will be one-half wavelength at the second harmonic. The reduced length may result in worse performance than allowing a small mismatch at the input frequency; it depends on the operating frequency and the magnitude of the junction capacitance. If nonlinear analysis is used, the stubs' length can be optimized easily on the computer.

4.4 SINGLY BALANCED DOUBLER: OUTPUT BALUN

4.4.1 Characteristics

As a singly balanced multiplier, this has nearly the same characteristics as the multiplier in Section 4.3. In this case, however, the balun is at the output, and this arrangement may make the circuit more practical. Like the previous circuit, this one rejects odd-harmonic outputs; the most important of these is the fundamental frequency. In many ways this circuit is the "dual" of the previous one.

This circuit has no significant advantages or disadvantages in conversion efficiency compared to the one in Section 4.3. Both are equally bad.

4.4.2 Description

Figure 4.10 shows the prototype circuit for this multiplier. Before you conclude that it simply is a reversed version of the circuit in Figure 4.7, note the diode polarity. It's reversed. Because of this, the dc diode current circulates in the diode-transformer loop, and as long as the input of the balun is a short-circuit at dc, no current return is needed. A current-return path for the RF still is needed. The center-tap of the transformer balun in Figure 4.10 provides this return, but microwave baluns generally have no analogous structure. Since it has a lower frequency, the RF current return in this circuit may be easier to implement. This circuit has the same output-current waveform as the input-balun multiplier shown in Figure 4.7.

The load resistance is electrically in series with the two diodes. Lacking any impedance transformation in the balun, each diode is terminated, at the second harmonic, with a resistance of $Z_L/2$. The diodes are in parallel with the input port, however, so each diode sees a source impedance of $2Z_s$, where Z_s is the source impedance. In Section 4.3 we noted that resistive multipliers require a load impedance approximately equal to the source impedance. Now, if $Z_L \approx Z_s$, it appears that we have a serious mismatch; one port or the other is four times its required value. This is much the same situation that we faced with the multiplier in Figure 4.9.

In this circuit the balun or transformer must provide an impedance transformation. The only trade-off is the balun bandwidth: when the balun is used to transform impedances, its bandwidth shrinks a bit. Still, bandwidths close to an octave are possible, and that's not bad, especially compared to varactor circuits.

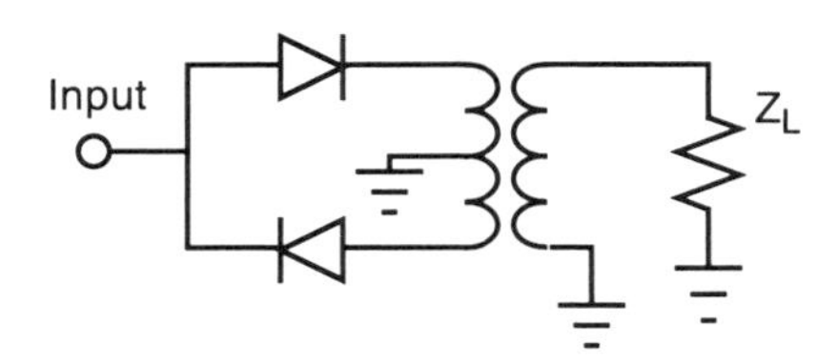

Figure 4.10 Balanced doubler using an output hybrid. Note that, compared to Figure 4.7, the diodes are reversed.

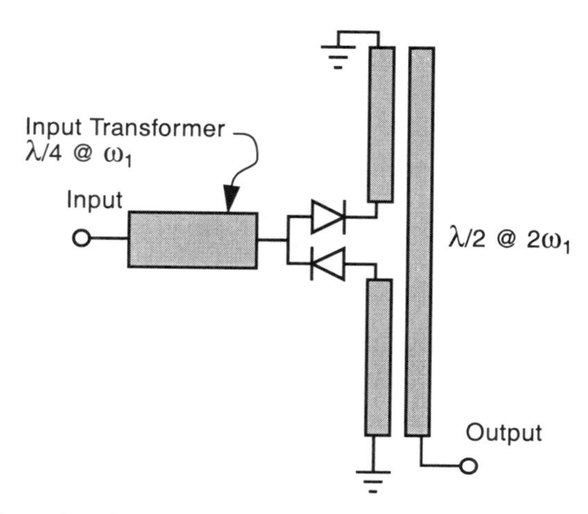

Figure 4.11 Singly balanced multiplier using an output balun. Although a two-strip coupled-line output balun is shown, a four-strip Lange-like structure may be needed.

4.4.3 Design

Figure 4.11 shows one of many possible circuits. It uses a Marchand balun for the output (see Section 3.6.3) and is good for about a 1.5:1 bandwidth. To minimize the even-mode impedance of the coupled-line balun, this multiplier is best realized on a thick, low-dielectric-constant substrate or a suspended substrate. On such substrates it may be necessary to use multiple coupled lines in the balun, as is done in a Lange coupler, to achieve the desired odd-mode impedance.

This design uses an input transformer and an impedance transformation in the output balun. These transformations are each only about 2:1, so the balun and the input transformer should be fairly broadband.

The fundamental-frequency current circulates in the balun. The balun should present a low impedance to this current, but in reality its inductance may not be negligible, especially near the high end of the input band. It may be necessary to tune the input slightly to compensate.

The multiplier's design process is as follows:

1. As before, create a single-diode prototype multiplier using (4.1) through (4.7) and the information in Section 4.2.3, with an input impedance around 50Ω.
2. Design the balun. This is easy. The odd-mode impedance is

$$Z_{0o} = \sqrt{\frac{Z_s Z_L}{2}} \tag{4.23}$$

where Z_s is the source impedance (50Ω) and Z_L is the load impedance of the single-diode prototype, found from (4.5). The even-mode impedance should be at least three times the odd-mode impedance, preferably higher. The length of each arm is $\lambda/4$ at the second harmonic.

3. Design the input transformer. Its impedance is

$$Z_0 = \sqrt{\frac{Z_s Z_{in}}{2}} \tag{4.24}$$

where Z_s is the 50Ω source impedance and Z_{in} is the input impedance of the single-diode prototype, from (4.3).

4. Using nonlinear analysis, add any tuning necessary to optimize the input VSWR.

5. Adjust the lengths of the balun arms slightly to optimize efficiency. *Be careful*: if the optimum balun lengths are much shorter than $\lambda/4$ at the output frequency, you probably have not optimized the input tuning, and you are compromising both the input VSWR and the output power by trying to adjust the balun length to optimize both.

4.4.4 Variations

Most of the circuits in Section 4.3 can be reversed and used as output-balun multipliers. The input and the output are interchanged, and one of the diodes must be reversed. Some other changes must be made; for example, the hybrid in Figure 4.8 must be tuned to the second harmonic, not the fundamental frequency. If you understand the basic principles of these circuits, the changes should be obvious.

4.4.5 Cautions

The most likely problem is that the fundamental-frequency inductance of the balun's arms will be too great to achieve a decent input VSWR. This type of balun becomes an open circuit to in-phase excitation as the frequency approaches the lower end of its passband. To prevent problems, the upper end of the input band must not be more than about 70% of the low-frequency end of the output band. This is one circuit that really should be analyzed on a computer.

4.5 DOUBLY BALANCED RESISTIVE FREQUENCY DOUBLER

4.5.1 Characteristics

One of the most irritating problems in designing singly balanced multipliers is the need for a ground return at either the fundamental frequency or the second harmonic. It would be nice if we didn't have to provide this return. Eliminating it would remove the last real impediment to achieving broad bandwidth, leaving us with a frequency multiplier whose bandwidth was limited only by the baluns.

This lofty goal can be accomplished by using a diode bridge and baluns at both the input and output frequencies. The fundamental configuration is shown in Figure 4.12. Such a multiplier is similar in many ways to a doubly balanced mixer: it isolates the input and the output completely and rejects odd-harmonic outputs. This geometry eliminates situations where the diodes are in parallel at the input and in series at the output or vice versa, making the ports easier to match. Finally, the number of diodes is doubled, so the input and output powers also are doubled, compared to a singly balanced multiplier. This may be helpful in achieving higher output power.

4.5.2 Description

A nice circuit for a doubly balanced multiplier is shown in Figure 4.13. This circuit uses two parallel-strip baluns on a suspended substrate. To equalize mode velocities in the balun, use a soft, low-dielectric-constant composite substrate.

This circuit can easily achieve a bandwidth of 3:1, and 4:1 or more is not unusual.

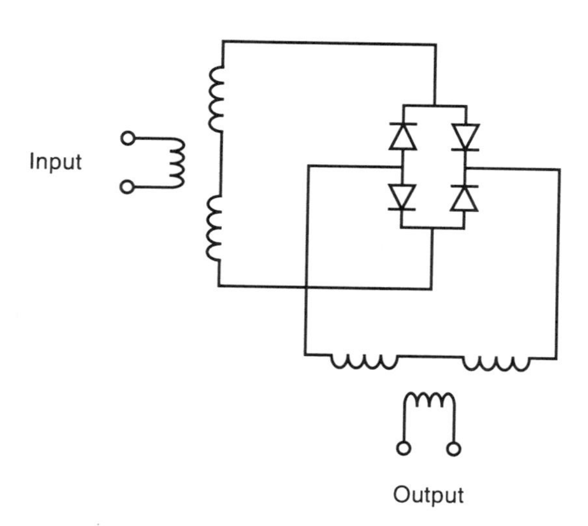

Figure 4.12 Basic circuit of a doubly balanced multiplier. The output currents circulate only in the diodes and the output transformer, while the fundamental currents circulate only in the diodes and the input transformer. Note the polarity of the diodes; if it is wrong, the multiplier won't work. Don't try to use a diode quad designed for mixers; use a so-called *bridge quad* instead.

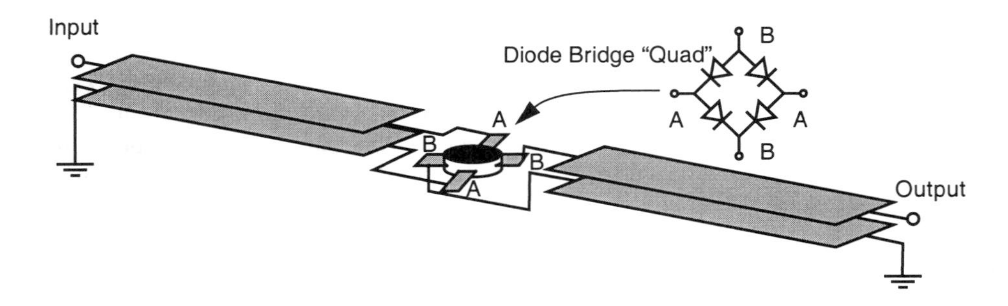

Figure 4.13 Doubly balanced frequency doubler: a simple, elegant circuit. The polarity of the diodes must be as shown.

4.5.3 Design

The process is as follows:

1. As with the singly balanced multiplier, design a single-diode multiplier as a prototype circuit. Use Equations (4.1) through (4.7) and Section 4.2.3. Keep the input impedance around 50Ω and be sure that the junction capacitance and capacitive parasitics are negligible at the highest output frequency.
2. Use a composite substrate having a dielectric constant below 3.0. It should be as thin as practical, to maximize the even-mode characteristic impedance. A 250-μm substrate is a good compromise between thickness and ease in handling.
3. Design the coupled lines of the balun to have an odd-mode characteristic impedance of $Z_{in}/2$, from (4.3), approximately 25Ω. The even-mode characteristic impedance should be at least 125Ω; the higher it is, the wider the bandwidth will be. The baluns should be at least about 30 degrees long, in terms of the even-mode phase velocity, at the low ends of their respective bands. They should be less than about 150 degrees long at the high end.
4. If necessary, a short, high-impedance line section can be inserted between the balun and the diode quad to tune out some of the diode capacitance. You may need to use nonlinear analysis to do this accurately.

4.5.4 Variations

These circuits work well in monolithic form when Marchand baluns are used. For an example of such a circuit, see [8].

4.5.5 Cautions

It's usually pretty easy to design these circuits and to make them work. There isn't much that can go wrong. The most likely error is to get the diode polarity wrong. If this happens, the circuit will do a very nice job of rejecting the second harmonic instead of the fundamental frequency.

4.6 VARACTOR FREQUENCY MULTIPLIERS

4.6.1 Characteristics

Varactor frequency multipliers have a well-deserved reputation for being troublesome. The fundamental problems are sensitivity and stability. Varactor circuits are sensitive, in that their performance varies dramatically with relatively small changes in circuit parameters. Tuning a varactor frequency multiplier can be a frustrating experience; it has been compared, probably accurately, to balancing a needle on its point. This high sensitivity makes monolithic varactor frequency multipliers impractical. Virtually all such circuits are hybrids.

It may seem strange that a varactor frequency multiplier can become unstable. After all, a varactor is a passive component. Surprisingly, however, a varactor can become active; under certain conditions, which invariably involve pumping it with a high-frequency sinusoid, it can generate a negative resistance. In the past, this principle has even been used to create negative-resistance varactor amplifiers, called *parametric amplifiers*.

Conversely, varactor frequency multipliers do have a significant advantage over the alternatives: noise. Varactors are virtually noiseless. This means that a varactor multiplier adds virtually no amplitude or phase noise to the signal being multiplied.[3] Additionally, they have much better efficiency than resistive multipliers, having conversion losses of only a few decibels, and operate at frequencies—millimeter waves—where active multipliers cannot be produced. Still, because of their finickiness, varactor multipliers probably are a last choice for a frequency multiplier; they are used mainly when none of the alternatives is acceptable.

Balanced varactor frequency multipliers are possible, although rare. Getting one diode to work properly is trouble enough; the advantages of a balanced multiplier rarely are worth the pain of getting two diodes to work together. Similarly, high-harmonic multipliers are more difficult to produce than doublers or triplers, and although such multipliers have been produced in the past, they rarely are designed today. For this reason we consider only doublers and triplers.

---

3. Varactor multipliers are not immune to the increase in phase noise caused by multiplication. All frequency multipliers increase phase noise by at least $20 \log(n)$ dB, where n is the harmonic number.

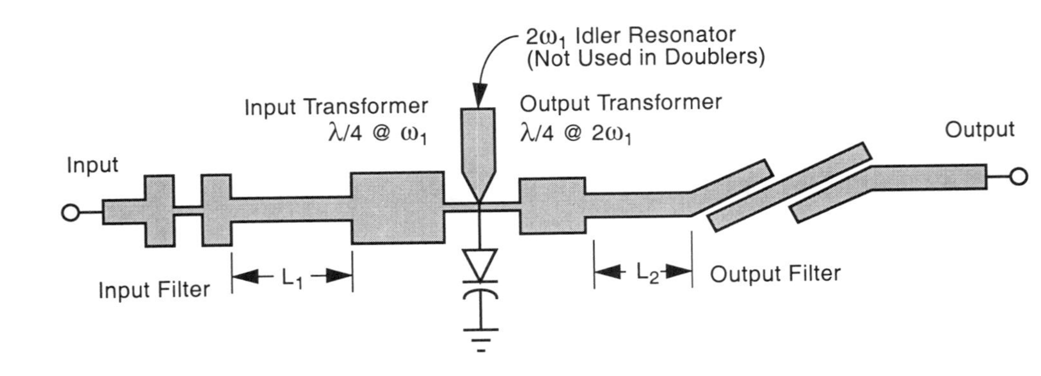

Figure 4.14 Varactor frequency multiplier. The lengths L_1 and L_2 are adjusted to make the input circuit look like an open circuit, at the diode, at the second harmonic. L_2 is adjusted similarly for the fundamental.

4.6.2 Description

Figure 4.14 shows a simple frequency multiplier. The input consists of a low-pass filter, transformer, bias circuit, and a length of transmission line. The output consists of a transformer, filter, and another transmission-line segment. Short high-impedance series lines are used as tuning inductors on the input and output. As with single-diode mixers, these circuits must be designed to provide proper source and load impedances to the diode and to prevent interaction between the input and output. If the circuit is a tripler, a second-harmonic idler resonator is needed.

One common and very satisfactory way to fabricate this circuit is to use a soft composite substrate and a packaged varactor. The varactor is mounted in a hole in the substrate, minimizing parasitics and providing a good heat sink. Remember that the varactor, although officially a reactive device, has series resistance, and most of the power lost in the multiplier is dissipated in this resistance. Good heat-sinking is essential.

4.6.3 Design

The design process is first to determine the input and output impedances of the varactor, design the filter/matching networks to provide those impedances, and finally to adjust the length of the input circuit's transmission-line sections to provide an open circuit to the diode at the second harmonic, and the output line to provide an open circuit at the fundamental.

Specifically, here is the process:

1. Select a diode. Use the information in Section 4.1.2 and Table 4.2 or 4.3. This is an iterative process: select parameters, substitute them into (4.11) through (4.14), and see if the performance is adequate. If not, change the parameters and repeat. This will give you the values of the input and output R-C models shown in Figure 4.3.
2. Include package parasitics in the diode model and calculate the diode-model impedances at the input and output frequencies. If this is a tripler, you must include the idler resonator as well. See Step 3.
3. A tripler requires an idler. This can be as simple as an open-circuit stub. Remember, the diode reactance found from S_{02}/S_{max} in Table 4.3 and the package parasitics are part of the idler, and the stub reactance must be adjusted accordingly. Unless the package parasitics are severe, the stub usually must be slightly inductive for the combination to be resonant. This stub will also affect the match at the input and output frequencies, so it must be included, along with other diode parasitics, in Step 2.
4. Design the input circuit:
   a. Use a standard approach for the low-pass filter; see [7] or any other good book on microwave filters. This filter should pass only the fundamental frequency and have at least 15-dB rejection at the second harmonic. For a tripler, its output impedance must be very low at both the second and third harmonics of the fundamental frequency.
   b. Design an input matching circuit consisting of a transformer and a short section of high-impedance line that matches the impedance of the complete diode model (Figure 4.3 plus package parasitics and idler stub) to the 50Ω source. This can be done on the computer or with a Smith chart.
   c. Compute the output impedance of the filter/matching structure, not including the diode. This is best done on the computer. Adjust the length of the 50Ω transmission-line segment between the filter and the matching circuit (L_1 in Figure 4.14) until the output impedance is an open circuit at the *output* frequency. This guarantees that the input circuit will not affect the output. If the filter is well designed, the length of this line will not affect the fundamental-frequency matching.
5. Design the output circuit in the same manner as the input circuit. The output filter provides virtually all the fundamental-frequency rejection, so it usually must be fairly complex. A simple structure, such as a stub, is rarely adequate. A coupled-line or comb-line filter is commonly used. Again, see [7] for help in designing the output filter.
6. Provide dc bias and a current return. (To retain clarity, these are not shown in Figure 4.14.) Since there is always some degree of rectified diode current, a

resistor can provide the required negative dc bias voltage; a voltage source is not needed.

7. Finally, optimize the circuit on the computer. A harmonic-balance analysis is best, but at least optimize the matching circuits using linear analysis.

4.6.4 Variations

This multiplier is trouble enough. Let's not worry about other ways to make one.

4.6.5 Cautions

Don't expect this circuit to work right out of the box. You will have to suffer a little to make it do what you want. Remember Thomas Edison's famous admonition that nothing works just to make you happy. This is a superlative example.

Designing frequency multipliers is partially a technological effort and partially a spiritual quest for personal growth. Try to think of it that way when you get really frustrated. And don't kick the dog.

4.7 STEP-RECOVERY-DIODE FREQUENCY MULTIPLIER

4.7.1 Characteristics

An SRD frequency multiplier is a reactive device, but it operates somewhat differently from a conventional varactor multiplier. SRDs are used primarily to generate high harmonics from a relatively low-frequency excitation. They also are used as fast pulse generators for sampling, radar, and similar applications. SRDs can generate large pulses a few tens of picoseconds long. They're fast. As with any reactive device, noise usually is relatively low, and the inevitable 20 log(n) degradation in phase noise invariably dominates any noise introduced by the device. Because SRD multipliers invariably are high-order, conversion loss can be high. Losses of a few to several tens of decibels, depending on the harmonic, should be expected.

High-order SRD multipliers, unlike varactor multipliers, do not require idlers. This is a very nice property; it is a consequence of the SRD's much stronger nonlinearity. The varactor's nonlinearity is essentially a depletion capacitance, while SRDs use the diode's diffusion capacitance for charge storage. An inevitable characteristic of a high-order multiplier is that the harmonics are closely spaced; for example, a tenth-harmonic multiplier has harmonics only 10% above and below the desired output, and these usually must be suppressed at least 20 dB. A high-order multiplier needs a pretty good output filter.

SRD multipliers, like varactors, are prone to instability. They also are sensitive to circuit parameters, and making one work requires a high tolerance for frustration.

4.7.2 Description

Because of the wide range of harmonic order used in SRD multipliers, it is a little difficult to identify a single circuit that can be used for a wide variety of applications. For this reason we have adopted the multiplier shown in Figure 4.15 as a "straw-man" circuit and hope that, once understood, it can be can modified appropriately for other uses. The circuit in Figure 4.15 is appropriate for input frequencies in the low microwave range, around 0.1–1 GHz, outputs between about 5 and 20 GHz, and an output up to perhaps the 10th harmonic. Above this frequency, SRD multipliers lose efficiency rapidly, and some other form of signal generation is likely to be more appropriate.

In Figure 4.15 we have included an output filter and a quarter-wave transmission line. This allows us to obtain a clean output at ω_n (4.18), which ideally will be equal to the output frequency, without affecting the pulse-formation process. Remember that the input impedance of a transmission line is equal to its characteristic impedance as long as there are no reflections, or until the reflected wave reaches the input. If the transmission line is $\lambda/4$ long at ω_n, the pulse "sees" a terminating impedance of Z_0, which is made equal to the load resistance. The pulse forms normally and propagates down the line. If the input of the filter is an open circuit in its stopband, the pulse is reflected and returns to the diode just as the diode switches off. The diode is now a short circuit, so the pulse is inverted and reflected, and the process repeats. The result is a damped sinusoid of voltage at the end of the transmission line. The narrowband filter then converts this damped sinusoid into a continuous sinusoid at the output frequency.

4.7.3 Design

The design consists of developing a pulse generator and attaching input and output filters to it. The input low-pass filter is essential; remember, in the design theory (Section 4.1.3) we assumed the source impedance to be zero at all frequencies except the fundamental; the input filter provides this termination. It also can be used as a matching circuit to transform the 50Ω source to the input impedance calculated in (4.20).

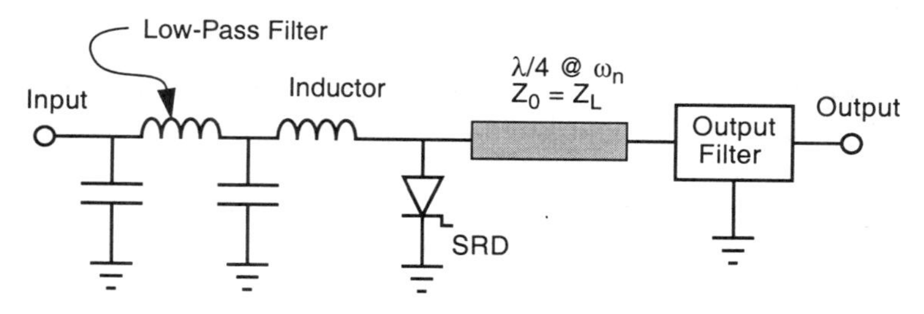

Figure 4.15 Basic SRD multiplier circuit.

The design process is to select a diode and to juggle (4.17) through (4.22) until a practical set of values emerges. Finally, the input filter and output structures are designed.

Diode Selection

SRDs have four important parameters; reverse breakdown voltage, V_b; reverse capacitance, C_d; minority-carrier lifetime, τ_n; and minimum transition time, τ_t. To begin with the obvious, the breakdown voltage must be greater than the maximum reverse voltage across the diode. The carrier lifetime must be long enough so little recombination occurs while the diode is charging; a good rule is

$$\tau_n \geq \frac{10}{f_{in}} \tag{4.25}$$

where f_{in} is the input frequency. There is no harm in making τ_n longer than this, except that the diode may be unnecessarily expensive. The transition time is the minimum switching time, which must be shorter than the output pulse length. Therefore,

$$\tau_t < T_p \tag{4.26}$$

where T_p is given by (4.17).

Virtually all the input power is dissipated in the diode. If the input power is high, the diode's temperature may rise considerably. The thermal resistance, θ_{jc}, and the input power, P_{in}, determine the temperature rise:

$$\Delta T = \theta_{jc} P_{in} \tag{4.27}$$

where ΔT is the difference in temperature between the diode junction and the mounting surface.

Multiplier Design

1. Select an appropriate value of the damping factor, ζ. Lacking any better insight, use $\zeta = 0.5$.
2. Use (4.18) through (4.22) to find practical values of L, C_d, and R_L that provide appropriate input impedance and power level. A constraint on this step is that ω_n should be equal to the output frequency.
3. Design the input low-pass filter, including impedance transformation if R_{in} differs from the source impedance. Reference [7] is perfect for this task.
4. Design an output transmission line and filter. The filter must be an open circuit in its stopband.

5. Put it all together. Expect to spend some time adjusting the multiplier manually.

4.7.4 Variations

Most alternatives are trivial variations of the circuit in Figure 4.15, usually different output filters or ways to couple to the diode. Occasionally the quarter-wave line is eliminated, although its function is implicitly retained by using a filter with an open-circuit stopband or some similar property.

4.7.5 Cautions

The most important concern in designing these creatures is stability. SRDs are prone to instability, perhaps more so than conventional varactor multipliers. While working on an SRD multiplier, you should monitor the output with a spectrum analyzer for signs of misbehavior. In the most common form of instability, the output breaks up into a triangular spectrum of closely spaced frequency components. Frequency-multiplier designers have invented a new verb to describe this phenomenon: *Christmas tree*, as in, "Man, I almost got that multiplier to work really well, and it suddenly started to Christmas tree on me." Learn the jargon, and you'll impress the management.

Instability is insidious. A multiplier that seems fine at room temperature might go into oscillation at higher or lower temperatures. Similarly, a multiplier that seems OK might break up when the power level increases or even decreases. You can never be too careful in making sure a multiplier is stable.

REFERENCES

[1] Page, Chester H., "Frequency Conversion with Positive Nonlinear Resistors," *J. National Bureau of Standards*, Vol. 56, No. 4, April 1956, p. 179.

[2] Maas, S. A., *Nonlinear Microwave Circuits*, New York: IEEE Press, 1997.

[3] Burckhardt, C. B., "Analysis of Varactor Frequency Multipliers for Arbitrary Capacitance Variation and Drive Level," *Bell System Technical J.*, Vol. 44, April, 1965, p. 675.

[4] Hamilton, S., and R. Hall, "Shunt-Mode Harmonic Generation Using Step-Recovery Diodes," *Microwave J.*, Vol. 10, No. 4, April 1967, p. 69.

[5] Hedderly, D. L., "An Analysis of a Circuit for the Generation of High-Order Harmonics Using an Ideal Nonlinear Capacitor," *IEEE Trans. on Electron Devices*, Vol. 9, 1962, p. 484.

[6] Ogawa, H., T. Hirota, and A. Minagawa, "Uni-Planar MIC Balanced Multiplier," *IEEE 1987 MTT-S Int. Microwave Symp. Digest*, p. 181.

[7] Matthaei, G., L. Young, and E. Jones, *Microwave Filters, Impedance-matching Networks, and Coupling Structures*, Norwood, MA: Artech House, 1980.

[8] Maas, S. A., and Y. Ryu, "A Broadband, Planar, Monolithic Resistive Frequency Doubler," *IEEE 1994 Microwave and Millimeter-Wave Monolithic Circuits Symp. Digest*, p. 443.

Chapter 5
Other Diode Applications

In this chapter we examine a few circuits that don't fit conveniently into any of the earlier topics. These include diode detectors and several kinds of modulators. An interesting property of many of these components is the complete lack of any published approach to their nonlinear design. We'll try to rectify that problem here.

Table 5.1 lists the circuits described in this chapter. The first two, square-law and envelope detectors, are different uses of what is largely the same component. The next, the double-sideband (DSB) modulator, is essentially just another use for a balanced mixer. The remaining two, single-sideband (SSB) and vector modulators, might be considered subsystems rather than components; nevertheless, there is more to these than just selecting a mixer or hooking together a few off-the-shelf components. Many aspects of the design of these components is surprisingly subtle. They are not as simple as they seem.

5.1 DIODE DETECTORS

There are two main applications for detectors: envelope detectors and square-law detectors. An *envelope detector* is used mostly for demodulating amplitude-modulated signals. It produces an output voltage that is proportional to the envelope of the signal. A *square-law detector* is essentially a power-measuring device. It produces an output voltage proportional to the power of the input signal (or, if you will, to the square of the input voltage; thus, the name). One of the most common applications for square-law detectors is in radiometers. These measure the power of broadband noise for noise measurement, radio astronomy, and radiometric imaging. Diode square-law detectors are also used in test instruments.

RF and microwave detectors invariably use Schottky-barrier diodes. *Pn*-junction diodes are useful only at very low frequencies. See Section 2.1.1.

Table 5.1 Components in This Chapter

Component	Characteristics	Typical Applications
Diode square-law detector	Output voltage is proportional to input power at low levels, input voltage at high levels. High input VSWR. Load resistance must be high.	Power measurement in test equipment and radiometers.
Diode envelope detector	Output voltage is proportional to the envelope of the input voltage waveform.	Demodulation of amplitude-modulated signals.
Double-sideband (DSB) modulator	Uses a doubly balanced mixer to generate DSB signals. Characteristics are largely those of the mixer.	Generation of double-sideband, suppressed-carrier signals.
Single-sideband (SSB) modulator	Uses a pair of doubly balanced mixers to generate SSB signals. Requires several hybrids and a 90-degree phase splitter in the modulating waveform. This can be difficult to realize.	Generation of single-sideband, suppressed-carrier signals.
Vector (I-Q) modulator	Generates a wide variety of digitally modulated signals. Requires two mixers and a 90-degree hybrid.	Modulation in a wide variety of digital communication systems.

5.1.1 Fundamental Properties

Both square-law and envelope detectors use essentially the same devices and circuits; the difference is in the RF power levels. At low levels, a detector operates as a square-law device; at higher levels, as an envelope detector.

Two basic detector circuits are shown in Figure 5.1. The series configuration, Figure 5.1(a), probably is the most common. The circuit is very simple; it is essentially a half-wave rectifier. C_b is a bypass capacitor and L is a current return inductor. The reactance of C_b is small at the lowest RF frequency, and the reactance of L is high, so the diode effectively is in parallel with the RF input port at high frequencies. At output (called *video*) frequencies, the diode effectively is in parallel with the output port, and the combination of R_v and C_b form an RC filter that rejects the high-frequency components. The inductor can be used as a tuning element to increase sensitivity by resonating the junction capacitance. If the input frequency is very high, it may be difficult to obtain an adequate bypass capacitor. In this case a transmission-line stub may be used.

Figure 5.1(b) shows a shunt configuration. It has no significant performance advantages over the series circuit but may be more practical in some cases; for example, certain types of diodes, especially pill-packaged ones, may be easier to mount in a shunt configuration. This circuit requires one extra component, the input dc blocking capacitor, C_c.

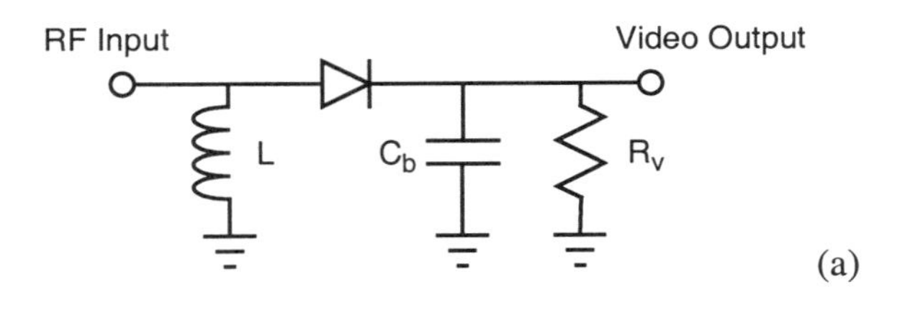

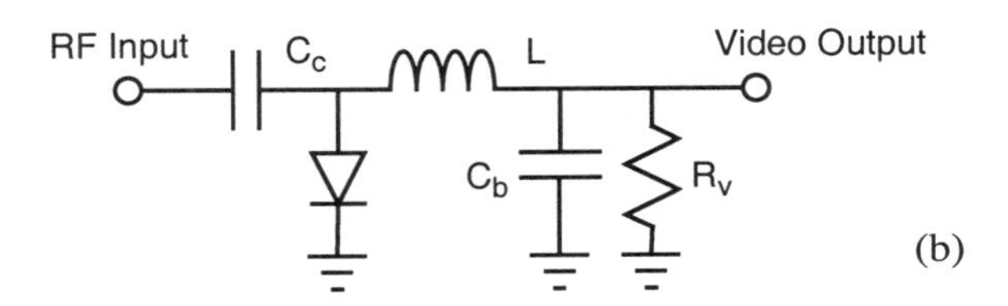

Figure 5.1 Two detector circuits. Circuit (a) is most frequently used; (b) may have practical advantages in some cases. The two circuits are virtually equivalent in RF performance. The video resistance, R_v, often is part of the external network, not part of the detector itself.

5.1.2 Detector Diodes

To function as a detector, the diode must have a low forward voltage drop and a high reverse resistance. Most Schottky diodes satisfy the latter requirement but not the former. A low forward voltage drop implies that I_s is quite high (see Section 2.1.2 for a discussion of the electrical characteristics of Schottky diodes). To achieve these characteristics, the diode must be heavily doped, which results in high junction capacitance and reverse leakage. Silicon diodes have much lower forward voltage drop than GaAs ones; therefore, virtually all detector diodes are silicon.

For microwave and millimeter-wave detectors a high junction capacitance may not be tolerable. To reduce the junction capacitance, conventional mixer Schottkys may be used, but dc bias is necessary. Providing dc bias to a detector usually is a nuisance; not only does it complicate the circuit, but drift in the bias voltage can cause changes in the detector's sensitivity. Furthermore, temperature changes cause the diode's I/V characteristic to change and result in a shift in the diode's dc bias current. This may be difficult to distinguish from a change in input level.

Most Schottky detector diodes use p material. Although p materials have much lower mobility than n materials, resulting in higher series resistance, they have lower $1/f$ noise. This is the dominant noise source in detector diodes and the main limitation to a detector's sensitivity. A detector's sensitivity usually is specified by a quantity called *tangential sensitivity*, a sloppy concept. It is discussed further in Section 5.1.3.

Most diode manufacturers provide special diodes for use as detectors. These are listed in catalogs as *zero-bias Schottky-barrier detector diodes.*

Finally, we recognize that a few detectors using point-contact silicon diodes are still in existence. (So are a few Model-T Fords!) Although such diodes have low barriers and may be sensitive, their mechanical fragility, cost, and limited frequency response make them unsuitable for new designs.

5.1.3 Square-Law Detection

Square-Law Characteristics

At low enough levels, virtually any resistive nonlinear device operates as a square-law detector. To understand this, consider the Taylor-series representation of the device's I/V characteristic (where V_b is the bias voltage and we have ignored the dc bias current):

$$i(v) = \left.\frac{dI}{dV}\right|_{V=V_b} v + \frac{1}{2}\left.\frac{d^2I}{dV^2}\right|_{V=V_b} v^2 + \frac{1}{6}\left.\frac{d^3I}{dV^3}\right|_{V=V_b} v^3 + \frac{1}{24}\left.\frac{d^4I}{dV^4}\right|_{V=V_b} v^4 + \dots \quad (5.1)$$

The dc component of the current comes primarily from the even-degree terms. As long as the later even-degree terms in the series are smaller than the first one, the device will operate as a square-law detector. In the case of a diode, we have

$$I(V) = I_s\left[\exp(\delta V) - 1\right] \quad (5.2)$$

where δ and I_s are constants (see Section 2.1.2). This gives

$$i(v) = I_s \exp(\delta V)\left(\delta v + \frac{1}{2}\delta^2 v^2 + \frac{1}{6}\delta^3 v^3 + \frac{1}{24}\delta^4 v^4 + \dots\right) \quad (5.3)$$

Because $\delta \sim 30$ in most detector diodes, the later terms are negligible only if v is small enough. To make the fourth-degree term small enough, relative to the quadratic term, we must have

$$\frac{1}{4}\delta^2 v^2 \ll 1 \quad (5.4)$$

This requirement limits the input level for good square-law response to about –30 dBm when the source impedance is 50Ω. It also implies that "soft" diodes, which have lower values of δ, have better square-law performance. Unfortunately, such diodes also have lower sensitivity.

Square-Law Sensitivity

The square-law sensitivity[1] of a square-law detector is defined as output voltage per input power. It is usually listed in units of millivolts per milliwatt. Sensitivity of 1,000–2,000 mV/mW is the norm, although at some frequencies, with careful design, sensitivities approaching 10,000 mV/mW can be achieved.

Tangential Signal Sensitivity

Tangential signal sensitivity (TSS), or more commonly *tangential sensitivity*, is a specification of the minimum signal that a square-law detector can handle. Unfortunately, it is not a well-defined concept.

The idea behind tangential sensitivity is this: we view the video noise from the detector on an oscilloscope. (A video amplifier usually is necessary to raise the detector's noise to a measurable level.) We then apply an RF pulse to the detector, and adjust the RF level until the positive peaks of the noise without the RF signal are equal to the negative peaks of the noise with the RF signal. The RF level that makes the noise peaks "tangent" in this way is the tangential sensitivity. Tangential sensitivities of microwave detectors usually are in the range of –55 dBm to –50 dBm.

The problem in this measurement is obvious: the peak value of a noise waveform, or any random process, cannot be defined. Furthermore, how do we know that the noise is from the detector, and not from the RF source, the video amplifier, or the oscilloscope? Clearly, this concept is valid only if (1) the noise of the video amplifier, RF source, and other instrumentation is small compared to the diode's noise; (2) the video bandwidth is specified; and (3) we use a more precisely defined quantity than the peak noise voltage. Fortunately (?), the $1/f$ noise in diodes is usually much greater than the thermal noise of the RF source or video amplifier, so in most detectors the diode's $1/f$ noise establishes the tangential sensitivity. However, if the video bandwidth is much less than the $1/f$ noise spectrum (as in a radiometer's detector) or is outside the $1/f$ spectrum (as in an envelope detector for broadband modulated signals), tangential sensitivity may not be a valid measure of the detector's sensitivity.

5.1.4 Envelope Detection

When a detector is operated at a high level, its output voltage is approximately proportional to the envelope of the input waveform. We say approximately, because at high levels the diode behaves like a short circuit, in forward conduction, with a small internal voltage drop. Thus, the output voltage is proportional to $|V_{RF}| - V_d$, where V_{RF} is the RF input voltage and V_d is the voltage drop, usually a few tenths of one volt. This is not precisely the envelope of V_{RF}. Clearly, V_d must be made as

---

1. Sorry about this term. Try not to confuse it with *tangential sensitivity*, an entirely different quantity.

small as possible, or at least very small compared to $|V_{RF}|$. Zero-bias Schottky detector diodes usually satisfy this requirement admirably.

5.2 SQUARE-LAW DETECTORS

5.2.1 Characteristics

A square-law detector produces an output voltage that is proportional to the input RF power. The square-law characteristic can be very accurate over a limited range. Typically, square-law detectors operate over an input power range of –55 dBm to about –30 dBm. Below this level, diode noise prevents their use, and above this level the square-law "linearity" (i.e., linearity in terms of output voltage per input power) starts to degrade. These detectors invariably have high, reactive input impedances.

5.2.2 Description

Virtually all square-law detectors are simple single-diode detectors. At lower frequencies, L and C_b may be lumped elements; at higher frequencies transmission-line stubs are used. A radial transmission-line stub often is a good choice for broadband bypassing. Even when transmission-line stubs are used for bypassing, it's a good idea to include a chip capacitor to prevent damage to the diode from static electrical discharges in handling or electrical transients in the circuit to which the detector is connected.

5.2.3 Design

Component Values

The design of square-law detectors is straightforward. The values of the components are selected according to the following criteria:

$$\begin{gathered}\frac{1}{(R_0 + R_j)C_b} \ll \omega_{r,\,min} \\ \frac{1}{(R_v \parallel R_j)C_b} \gg \omega_{v,\,max} \\ \frac{R_0 \parallel R_j}{L} \ll \omega_{r,\,min} \\ \frac{(R_v + R_j)}{L} \gg \omega_{v,\,max}\end{gathered} \tag{5.5}$$

where $\parallel$ indicates the parallel combination of the resistances, R_0 is the RF source resistance, $\omega_{r,min}$ is the minimum RF input frequency, $\omega_{v,min}$ is the minimum video output frequency, and the other ω terms follow similarly. R_j is the diode's junction resistance,

$$R_j = \frac{1}{\delta I_s \exp(\delta V_b)} \tag{5.6}$$

which is on the order of several thousand ohms, invariably much greater than the RF source impedance. The other quantities are as shown in Figure 5.1. Most square-law detectors produce a slowly varying output; $\omega_{v,max} / 2\pi$ usually is no more than a few Hz. To achieve good sensitivity, the video resistance must be quite high, usually on the order of 100 KΩ. Normally $R_v >> R_j$. You may have to experiment with R_v somewhat to achieve the widest possible square-law range.

Diode Selection

The tricky part of detector design is the selection of the diode. As with mixers, we want to be sure that the series resistance and junction capacitance do not reduce the sensitivity. We need

$$\frac{1}{(R_0 + R_s)C_j} \gg \omega_{r,\,max} \tag{5.7}$$

and if the inductor L is used to tune the diode, the requirement is loosened a bit to

$$\frac{1}{R_s C_j} \gg \omega_{r,\,max} \tag{5.8}$$

where R_s is the diode's series resistance and C_j is its junction capacitance. If these requirements are met, the input impedance of the detector is simply R_j. The lowest possible value of R_j is several hundred ohms, so an input match to a Schottky-diode detector is rarely possible.

The diode's value of I_s must be selected to optimize the detector's sensitivity. There is an optimum value of I_s. If I_s is too low, both the sensitivity and the square-law characteristic will be poor. If it is too great, the low junction resistance, R_j, which is in parallel with the video output port, will reduce sensitivity. The sensitiivity analysis in the next section can be used to make this trade-off.

Sensitivity

Volterra methods [1] can be used to derive an accurate expression for the square-law sensitivity of the detector in Figure 5.1(a). The RF junction voltage, v_j, is

$$v_j = \frac{\sqrt{8P_{av}R_0}R_jLs}{(R_0+R_s)R_jLC_js^2+\left[(R_0+R_s+R_j)L+R_0R_sR_jC_j\right]s+(R_j+R_s)R_0} \quad (5.9)$$

where R_0 is the RF source impedance, P_{av} is the available power of the source, and $s = j\omega$. The video output voltage is

$$V_o = \frac{0.5\delta}{R_j}|v_j|^2\frac{R_vR_j}{R_v+R_j} \quad (5.10)$$

Equations (5.9) and (5.10) were derived under the assumption that the input is sinusoidal and that the diode is unbiased. They are also valid when the RF input signal is broadband noise.

5.2.4 Cautions

In (5.6) we implicitly left the door open to the use of dc bias in a detector diode. Unless it is essential for sensitivity, dc bias should be avoided, because drift in the bias voltage or changes in the diode's characteristics with temperature can cause drift in the output voltage.

5.2.5 Variations

The *backward diode*, sometimes called the *back diode*, often is used instead of a Schottky diode. A backward diode is a type of diode in which the reverse-bias current actually is greater than the forward-bias current. It does this through the magic of electron tunneling; the diode does not break down.

We mentioned earlier that a difficulty in making detectors with Schottky diodes is the need for a very high value of I_s, or to put it another way, a low knee voltage in the diode's I/V characteristic. This is not a problem in back diodes; the knee of the reverse I/V characteristic is very low. Furthermore, the strength of the second-degree nonlinearity is much greater in back diodes than in Schottkys, so the sensitivity is better, and the square-law characteristic has a wider range. Unfortunately, the parasitic capacitances generally are greater.

5.3 ENVELOPE DETECTORS

5.3.1 Characteristics

An envelope detector produces an output voltage proportional to the voltage magnitude of the RF input signal. The circuits in Figure 5.1 operate as envelope detectors quite nicely, and the criteria of (5.5) are valid for their design. When those conditions are met, the output voltage is

$$V_o = (|v_r| - V_d)\frac{R_v}{R_0 + R_s + R_v} \tag{5.11}$$

where V_d is the voltage drop across the diode junction and v_r is the open-circuit RF voltage of the source. As long as V_d is small, this is a faithful reproduction of the envelope of the RF signal.

The main difference between envelope detectors and square-law detectors is input level. In most diodes square-law operation requires an input level below about −30 dBm; much higher levels (at least −10 dBm) are used for envelope detection. Between these levels, you get either poor square-law detection or distorted envelope detection. Take your pick.

5.3.2 Design

Equation (5.5) applies to envelope detectors as well as square-law detectors. The selection of a diode is not as critical as with a square-law detector, however; the main requirements are that the junction capacitance be appropriate, which is essentially what the criteria in (5.5) imply, and that the internal voltage drop, V_d, be small. The latter requirement is satisfied by keeping the diode's I_s parameter small, preferably by using a zero-bias Schottky detector diode.

5.3.3 Variations

Envelope detectors can be realized as full-wave rectifiers. The basic circuit is shown in Figure 5.2. An advantage of this configuration is the rejection of fundamental-frequency output-current components, so filtering is simpler. The required R_v-C_b time constant is only half that given in (5.5). Additionally, the transformer can increase the source voltage applied to each diode, thereby improving sensitivity and reducing distortion.

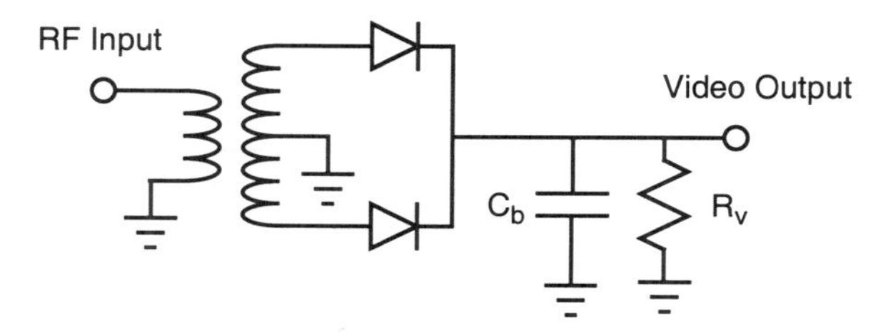

Figure 5.2 A balanced, full-wave envelope detector. In this configuration, each diode is driven by the output impedance of half the transformer's secondary winding. Sensitivity can be increased by the use of a step-up transformer.

5.3.4 Cautions

Envelope detectors are prone to an insidious form of distortion caused by the difference in charging and discharging rates of the R_v-C_b combination. Imagine that the signal's envelope is a large pulse. The detector diode is forward biased and C_b charges rapidly through R_0, as illustrated in Figure 5.3(a). After the pulse, however, the voltage on C_b is high, and the only path for it to discharge is through R_v. In virtually all detectors, $R_v >> R_0$, so the time constant for the discharge is much longer than the time constant for charging. As a result, the detector does not follow the peak well. The video output voltage has a fast rise but falls very slowly. The solution to this problem is to limit the values of R_v and C_b.

5.4 DOUBLE-SIDEBAND (DSB) MODULATORS

5.4.1 Characteristics

Many communication systems use double-sideband, suppressed-carrier amplitude modulation (DSB). The DSB signal is expressed simply as

$$V_{RF}(t) = m(t)V\cos(\omega t) \tag{5.12}$$

where V_{RF} is the RF signal, $m(t)$ is the modulating waveform, often audio, and V is the amplitude of the wave when $m(t) = 1$. The RF signal is simply the product of the modulating waveform $m(t)$ and a carrier, $V\cos(\omega t)$.

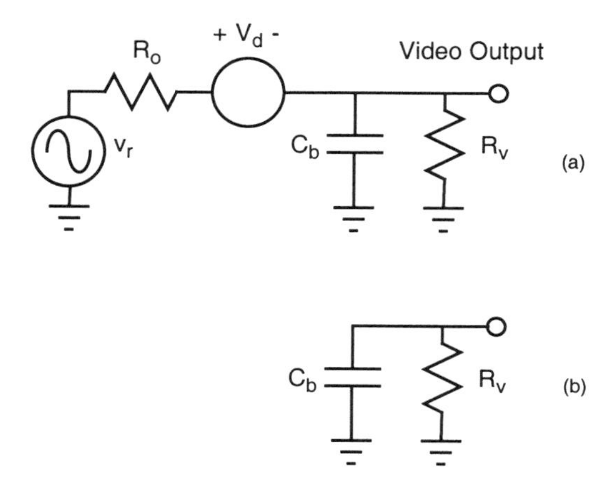

Figure 5.3 Circuits illustrating peak distortion in the envelope detector. Circuit (a) represents the detector when the diode is conducting on envelope peaks; (b) represents the detector after the peak. If $R_v >> R_0$, recovery from the peak is very slow.

Mixers fundamentally are multipliers, so some type of mixer should be ideal for this purpose. In this case, the modulating waveform is a baseband signal and the carrier is a high-frequency one, so the logical approach is to use the IF input for $m(t)$, the LO input for the carrier, and the RF port for the modulated output waveform. From (5.12) we can see, however, that not any mixer will do. The mixer must have a number of important characteristics:

- If the modulating signal extends to baseband, the mixer's IF bandwidth also must do so.
- To prevent the carrier from leaking into the output, the mixer must have high LO-to-RF isolation.
- The mixer must have low distortion.

These requirements usually dictate the use of a doubly balanced mixer.

5.4.2 Description

DSB modulators usually use one of two types of mixers: an active Gilbert-cell mixer (described in Section 6.7) or a diode ring mixer. Because of its high LO-to-RF isolation, high impedance levels, ability to handle large signal voltages without distortion, and easy monolithic integration, the bipolar-transistor Gilbert mixer probably is preferable. Diode modulators have the advantage of operation at much higher frequencies, and their distortion and LO-to-RF isolation may be adequate if they are treated carefully.

5.4.3 Design

Design largely consists of selecting an appropriate mixer and operating it properly. Some considerations are presented here.

Diode Mixers

Doubly balanced ring mixers with transformer baluns frequently are used as DSB modulators. They offer very good LO-to-RF isolation, broad bandwidth, and low distortion, but they are limited in frequency to a few hundred megahertz. At higher frequencies the microwave ring mixers described in Section 3.4 and following sections may be used, although some of these circuits may not have adequate LO-to-RF isolation for use as a modulator.

To minimize distortion, the modulating signal must be kept 10 to 20 dB below the carrier. This results in a relatively weak output, and a fair amount of amplification may be needed to raise it to a useful level. Minimizing distortion also requires low-VSWR, broadband terminations at all ports.

When used as a modulator, a mixer is operated as an upconverter. The diodes' junction capacitances cause diode mixers to have unusual characteristics when operated as upconverters. These characteristics include (1) increased distortion, compared to downconverter operation, and (2) asymmetrical sidebands in the output waveform. To avoid such problems, use relatively small diodes having low junction capacitance.

Active MESFET Mixers

Because of their inherently low LO-to-RF isolation, unbalanced active MESFET mixers are rarely a good choice for DSB modulators. Singly or doubly balanced mixers may be adequate, but their need for baluns complicates the circuit and makes it large. Unlike bipolar mixers, doubly balanced MESFET mixers do not have good LO rejection, and using such mixers without a balun usually degrades the LO-to-RF isolation.

BJT Mixers

Virtually all BJT mixers use Gilbert cells (see Section 6.7). A Gilbert cell is an analog multiplier whose structure is in many ways similar to an interconnection of differential amplifiers. They are usually realized as integrated circuits.

Modern BJT IC technology can produce transistors that have very closely matched characteristics. Mixers using such transistors have very good LO-to-RF isolation. They also have low distortion. Furthermore, the high even-mode rejection of the differentially connected BJTs frequently allows such modulators to be used without baluns.

Gilbert-cell mixers using BJTs operate at frequencies up to a few hundred megahertz. At frequencies up to about 20 GHz, HBTs can be used. For higher frequencies the modulated waveform can be generated at a lower frequency and upconverted.

5.4.4 Variations

Gilbert-like mixers can be realized in MOSFET technology as well as BJT. These should have many of the desirable characteristics of their BJT analogs.

5.4.5 Cautions

The most likely problems one might encounter with DSB modulators are carrier leak-through from poor LO-to-RF isolation in the mixer, and, especially in diode modulators, distortion. Such problems are easily avoided by careful selection of components and signal levels.

5.5 SINGLE-SIDEBAND (SSB) MODULATORS

5.5.1 Characteristics

An SSB modulator generates a single-sideband, suppressed-carrier modulated signal. As with the DSB modulator, the most difficult task is to eliminate the unwanted outputs which in this case are both the carrier and the unwanted sideband. The required suppression may be very great; even with 20 or 30 dB suppression of the unwanted sideband, it still may be great enough to cause interference to receivers near the transmitter. As with the DSB modulator, we also need to minimize distortion.

5.5.2 Description

A simple (and probably obsolete) way to generate such a signal is to generate a DSB signal and to filter out one of the sidebands. This requires large and relatively expensive filters, and, worst of all, must be done at a frequency low enough that the unwanted sideband can be filtered effectively. Such SSB generators rarely operate at frequencies above 500 KHz.

A better way is to use the circuit shown in Figure 5.4, which consists of two DSB modulators, a hybrid, and a 90-degree phase splitter for the modulating signal. This approach has a number of advantages over the filter method: it is smaller, less expensive, and can be integrated easily. The only real disadvantage is that the 90-degree phase splitter is theoretically impossible to realize.

Why is it impossible? An ideal, 90-degree phase splitter performs something known as a Hilbert transformation. A Hilbert-transforming filter is a classic example of an unrealizable component. It just can't be done. Fortunately, however, it is possible to approximate such a filter over a limited bandwidth. This usually is done by making two filters whose transfer functions' phases differ by 90 degrees.

The difficult part of the design of an SSB modulator is realizing this phase splitter. Routines for designing such circuits are standard fare for filter-design software. One such example is in [2].

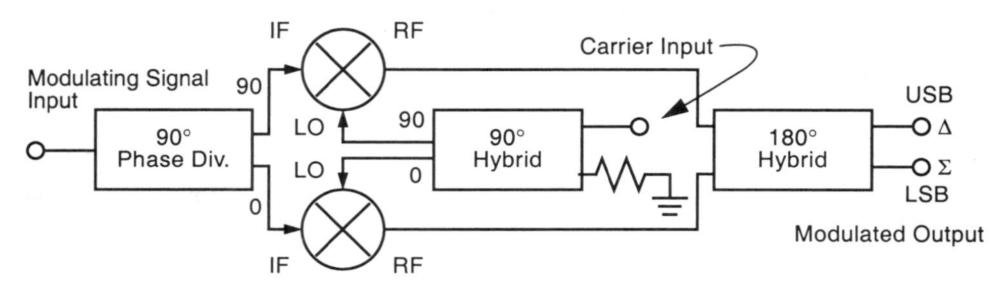

Figure 5.4 SSB modulator. Both an upper- and a lower-sideband signal are generated; the unwanted output port is simply terminated.

5.5.3 Design

The mixers in Figure 5.4 are in fact DSB modulators, so the considerations for their selection and operation are essentially those in Section 5.4.3. The critical characteristic in SSB modulators is the overall gain and phase balance of the mixers and hybrids. It is essential that the time delay through the two paths in the circuit of Figure 5.4 be identical; otherwise, the unwanted-sideband suppression will suffer.

Monolithic integrated circuits have a great advantage in equalizing the phase and the amplitude. The mixers can be made virtually identical, so the paths can be equalized precisely. If the modulator is fabricated from individual, discrete components, these characteristics are much more difficult to achieve. If individual components must be used, it is best to mount them precisely in a single housing and to use microstrip or stripline interconnections. Probably the worst approach is to interconnect the components via coaxial lines and connectors.

The sideband rejection can be predicted from the amplitude and phase errors as follows:

$$R_I = -10\log\left[\frac{1 - 2\sqrt{G}\cos(\theta) + G}{1 + 2\sqrt{G}\cos(\theta) + G}\right] \tag{5.13}$$

G is the gain imbalance (a factor, not decibels) and θ is the phase imbalance. Achieving 20 dB of unwanted sideband rejection requires, for example, that the path gains and phases be equalized within 1 dB and 10 degrees, respectively. 30 dB rejection requires 0.3 dB and 3 degrees. Care to try for 40?

5.5.4 Variations

In an integrated circuit the carrier's 90-degree hybrid can be difficult to realize. A useful trick is to shift the phase by splitting the carrier signal, shifting the phase of one of the resulting signals with an RC network, and finally passing both through saturated amplifiers to equalize the amplitudes.

A summing junction can be used in place of the 180-degree hybrid. When this is done, interchanging the outputs of either of the 90-degree hybrids will provide a different sideband.

5.5.5 Cautions

It is possible to create SSB modulators having different configurations from the one shown in Figure 5.4. Unfortunately, the 90-degree phase split in the modulating signal is always necessary. Don't believe anyone who tells you otherwise!

5.6 I-Q MODULATORS

5.6.1 Characteristics

The *I* and *Q* stand for *in-phase* and *quadrature-phase*. Sometimes called *vector modulators*, these circuits are used for digital modulation.

The idea behind this type of modulator is fairly simple. Many types of digital modulation involve shifting the phase and sometimes the amplitude of a sinusoidal signal. Any such signal can be represented by a vector whose amplitude and phase vary from bit to bit. Two orthogonal basis vectors span this space. Thus, by setting the amplitude of each of two vectors or, equivalently, the amplitudes of a 0 and 90-degree sinusoids, we can create a signal having the desired amplitude and phase for any bit.

In digital communication systems, noise causes bit errors; a noise fluctuation during any one bit period can change its amplitude or phase enough to cause it to be mistaken for another signal. Clearly, if the phase or amplitude of the signal are not precisely correct to begin with, bit errors will be more likely. Furthermore, the passbands of digital modulators usually must be very flat. Again, distortion in the frequency domain results in phase and amplitude shifts in the time domain.

Since the precision of the modulator is critical to the performance of the system, the amplitude, phase, and passband specifications for digital modulators often are stringent. As with an SSB modulator, monolithic technology can be a great help in keeping phases and amplitudes equal. The cautions in Section 5.5.3 apply here as well.

5.6.2 Description

The basic circuit of an I-Q modulator is shown in Figure 5.5. Two separate data streams are applied to the IF ports of the mixers. The LO is split in phase by 90 degrees. The output of the upper mixer becomes the quadrature component; the output of the lower is the in-phase. The two outputs are simply added.

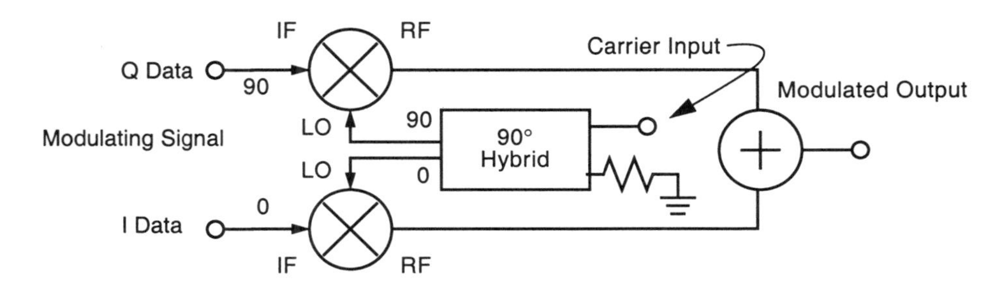

Figure 5.5 I-Q modulator.

5.6.3 Design

Design of this mixer is similar to the design of the SSB modulator. It is essential that the levels of the I and Q data streams do not saturate the mixer. Even if the streams carry no amplitude information, as with quadrature-phase modulation, saturating the mixer will result in phase shifts that degrade the phase accuracy of the modulated waveform.

The selection criteria for the mixers is much the same as in Section 5.4.3. Good passband flatness and very low LO leakage are essential in such mixers. Most doubly balanced mixers exhibit good phase-shift characteristics; if the LO signals at the mixers differ by 90 degrees, so will the outputs. Unlike the data streams, the LO levels should be fairly high, so the diodes are switched soundly and distortion is minimized.

5.6.4 Variations

In monolithic circuits, the 90-degree hybrid may be difficult to realize. The trick described in Section 5.5.4 for realizing this phase split may be useful here as well.

Some digital communication systems use biphase modulation, in which the modulated signal has either a 0- or a 180-degree phase. Such systems do not require I-Q modulators, since only a single basis vector is needed to span the space of all possible signal vectors. In such systems the digital data are applied to the IF port of a doubly balanced mixer at a high level, so the digital signal switches the diodes. The carrier is applied at a lower level to the LO port.

REFERENCES

[1] Maas, S. A., *Nonlinear Microwave Circuits*, Norwood MA: Artech House, 1988.

[2] Ellis, M. G., *PC Filter*, Norwood MA: Artech House, 1995.

Chapter 6
Active Mixers

In the early days of GaAs MESFET technology, we all expected FET mixers eventually to supplant diode mixers. This hasn't happened and probably never will. Active mixers—both bipolar and FET—are not uniformly superior to diode mixers. Their greatest advantages are conversion gain and compatibility with monolithic processes; their greatest disadvantage is the clumsiness of some of the balanced circuits.

Diode and active mixers are very different creatures, so it is difficult to compare their performance. The noise figures of the best FET mixers are only modestly better than those of diode mixers, but the FET mixers' conversion gain improves the overall noise figure and reduces the number of stages in a system. Similarly, input intercept points (IP) of FET mixers are often worse, but their gain reduces the number of stages of input amplification. The reduction of input-stage gain may improve a system's dynamic range.

As with the single-diode mixer, the single-FET mixer is a prototype for balanced circuits. Therefore, even if your only interest is in balanced mixers, it is worth the effort to read Sections 6.1 and 6.2 before embarking on a balanced-mixer design.

The circuits examined in this chapter are listed in Table 6.1. Many variations are possible. The circuits in Table 6.1 represent a selection of practical circuits that can be used as starting points in a design. Although most circuits use MESFETs, MOSFETs or JFETs can also be used in appropriate frequency ranges. The Gilbert mixer can be realized with either BJTs or HBTs.

6.1 ACTIVE MIXER THEORY

In Chapter 3 we saw that a large LO voltage applied to a diode caused its junction conductance to vary and that frequency conversion in a diode mixer is caused by this time-varying conductance. In an active mixer we apply the LO voltage between a transistor's gate and source (or base and emitter) terminals and use the LO to

Table 6.1 Mixers Described in This Chapter

Mixer Type	Characteristics	Typical Applications
Single-device, single-gate FET mixer	Simple, inexpensive. Provides conversion gain, low noise, low distortion. LO-to-IF isolation may be poor. Best for moderate-bandwidth applications.	Downconverter in receivers. The need for an LO-RF diplexer and IF filters limits applications.
Singe-device, dual-gate FET mixer	Good LO-to-RF isolation without filters. Gain and noise figure are somewhat worse than single-gate mixers; distortion is about the same. Moderate bandwidth.	A nice mixer for low-cost integrated circuits, especially commercial applications.
Singly balanced FET mixer	Essentially the same characteristics as a single-device mixer, but good LO-IF and LO-RF isolation without a diplexer. 3-dB higher LO power, 3-dB better IP. May require baluns.	High-performance IC applications where the number of FETs and the size of the baluns is acceptable. The large number of interconnections makes this a poor choice for hybrid circuits.
Doubly balanced MOSFET mixer	Essentially the same characteristics as a single-device mixer, but has all the benefits of a doubly balanced circuit (see Chapter 3). 6 dB higher LO power, 6 dB better IP. May require baluns.	Essentially the same applications as the singly balanced mixer, but where improved intermodulation rejection justifies the extra complexity and LO power.
Gilbert-cell BJT/HBT mixer	A doubly balanced bipolar mixer. Often works well without baluns. Can be used as a "linear" analog multiplier.	Modulators, signal processing as well as mixing.

vary its transconductance. Frequency conversion is caused by this time-varying transconductance.

Transistors have a number of nonlinearities that can be exploited for frequency conversion. In the past, transistor mixers (you may detect a reluctance to call them *active* mixers) have used a FET's drain-to-source resistance or gate-to-channel junction for mixing. These just don't work as well as transconductance mixers, and operating the device in this manner may compromise its reliability. For this reason, in this chapter we concern ourselves only with transconductance mixer circuits. Furthermore, to keep things as concrete as possible, in this section we consider only

FETs. We restrict our scope for two reasons: the great majority of active RF and microwave mixers use FETs, and, in any case, the extension of the theory to bipolar devices is straightforward.

6.1.1 Transconductance Mixers

Our model for a simplified FET transconductance mixer is shown in Figure 6.1. In this mixer, we apply both the RF signal and the LO voltage to the gate. The LO voltage creates the time-varying transconductance, and the RF mixes with it.

The design of an active mixer is based on a single idea: maximize the magnitude of the fundamental-frequency component of the transconductance. The reason is similar to the reasoning in Section 3.1.1. This results in the greatest conversion efficiency and the lowest noise figure.

The small-signal drain current, $i_d(t)$, is simply

$$i_d(t) = G_m(t)\, v_g(t) \tag{6.1}$$

where $G_m(t)$ is the transconductance waveform and $v_g(t)$ is the gate voltage. In general $G_m(t)$ is periodic but not sinusoidal, so we can express it as a Fourier series,

$$G_m(t) = G_{m0} + G_{m1}\cos(\omega_p t) + G_{m2}\cos(2\omega_p t) + G_{m3}\cos(3\omega_p t) + \ldots \tag{6.2}$$

where ω_p is the LO frequency and, for simplicity, we have omitted phase terms in each component. (If the conductance pulses are symmetrical, which is usually the case, and the zero point of the time axis is chosen properly, there will be no phase terms, anyway.) If we substitute the RF signal, $v_g(t) = V_s \cos(\omega_s t)$, and (6.2) into (6.3), the drain current becomes

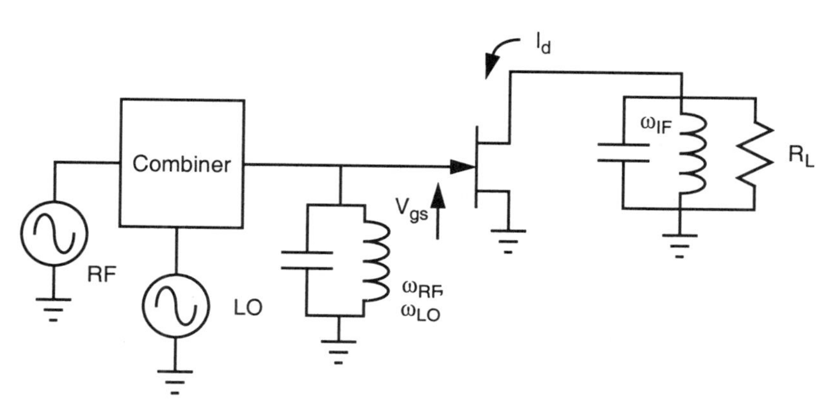

Figure 6.1 Idealized FET transconductance mixer. The FET has no parasitics and the bias sources are not shown. The combiner usually is a filter diplexer or, occasionally, a directional coupler. The output tuned circuit short-circuits the drain at all frequencies except the IF, and the input tuned circuit shorts the gate at all frequencies except the RF and LO.

$$i_d(t) = G_{m0} V_s \cos(\omega_s t) + \frac{G_{m1} V_s}{2}\left[\cos[(\omega_s - \omega_p)t] + \cos[(\omega_s + \omega_p)t]\right]$$
$$+ \frac{G_{m2} V_s}{2}\left[\cos[(\omega_s - 2\omega_p)t] + \cos[(\omega_s + 2\omega_p)t]\right] + \ldots \quad (6.3)$$

Our desired output is the $\omega_s - \omega_p$ component of the drain current; the rest is excess baggage and must be eliminated. (We'll say a little more later about the best way to do this.) *The only component of the conductance waveform that affects our desired output is* G_{m1}.[1] Thus, to maximize the IF output current, we must maximize $G_m(t)$, the fundamental-frequency component of the transconductance waveform, $G_m(t)$.

In all FETs, including HEMTs, the transconductance is maximum when the device is operated in current saturation. In this case the transconductance can be varied from zero (at pinch-off) to some maximum value. In MESFETs and other junction FETs, the maximum transconductance occurs at maximum channel current; in HEMTs, the transconductance peaks at a gate voltage somewhat below that which provides maximum channel current. Clearly, the LO must vary the transconductance between these two extremes. But how? Should we bias the gate halfway between maximum and minimum and have a sinusoidal $G_m(t)$ waveform? Or should we bias it at or below pinch-off and have a pulsed sinusoidal waveform? These options are illustrated in Figure 6.2.

A few minutes with a book of Fourier-series expressions provides the answer: if the transconductance is idealized as a linear function of gate voltage, and enough LO power is available to drive the FET to maximum transconductance, biasing the FET at pinch-off provides the greatest $G_m(t)$. In real devices, the optimum gate bias may not be precisely at pinch-off, but it will be close.

At this point, we have generated several rules for operating transconductance mixers:

1. Bias the device at pinch-off.
2. Drive the gate with enough LO power so the peak gate voltage provides maximum transconductance.
3. Keep the FET in current saturation throughout the LO cycle.

The methods for achieving these goals should be clear:

1. Conjugate-match the gate at the LO frequency.
2. Provide enough LO power.
3. Short-circuit the drain at all LO harmonics, including the fundamental. This will "pin" the drain voltage at its dc value, keeping the device in current saturation throughout the LO cycle. It also will prevent LO leakage into the IF circuit.

---

1. To be honest, the other components of $G_m(t)$ do have some effect; a less approximate analysis is necessary to show this. However, G_{m1} clearly is dominant.

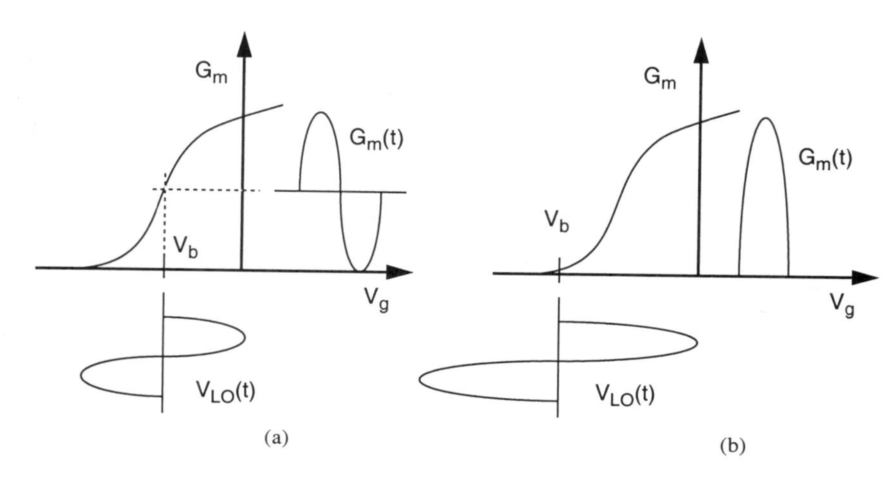

Figure 6.2 For high gain and low noise, the gate bias must be chosen to provide the greatest fundamental-frequency component of the transconductance waveform. In (a), the FET is biased in the active region; in (b), near pinch-off. The latter provides optimum performance.

The rest of the design is relatively straightforward; only a few subtleties need be mentioned. We consider these below.

RF Input Matching

To maximize conversion efficiency, conjugate-match the gate at the RF. There appears to be little value in mismatching the input to improve the noise figure of an active mixer, as is done in a low-noise FET amplifier. There are a couple of good reasons why this is true. First, in an amplifier, the input mismatch is a trade-off between the need for gain, which dictates a conjugate input match, and low noise, in which the source impedance is chosen to force the partially correlated gate and drain noise currents to cancel. The gain of a FET mixer inherently is much lower than an amplifier, so the trade-off of noise against gain must be biased more toward gain. Furthermore, the correlation between gate and drain noise currents in a mixer is much weaker, so less cancellation is possible.

LO Input Matching

If the drain is effectively short-circuited at the LO frequency, the input LO impedance, Z_{in}, is approximately

$$Z_{in} = R_s + R_i + R_g + \frac{1}{\omega_p C_{gs}} \tag{6.4}$$

where ω_p is the LO frequency and the other terms are elements of the FET's equivalent circuit, defined in Figure 2.18. Ideally, the LO input should be conjugate matched. Often, however, this is impossible in practice. At low frequencies, the input Q of a FET is very high and is virtually impossible to match over even a modest bandwidth. Even at higher frequencies it may be very difficult to match the input over both the LO and RF bands. In this case, unless LO power is scarce, it probably is best to match the RF and suffer some reflection loss at the LO.

At low frequencies it often is adequate to put an inductor in series with the gate to resonate C_{gs} and to adjust the source resistance (with a transformer) to achieve a reasonable bandwidth. Of course, more complex matching circuits are possible, but rarely do they have a significant advantage over the simple, practical approach described in Section 6.2.3.

IF Output Matching

In a downconverting FET mixer, the IF output impedance is very high. The IF output effectively is a current source in parallel with a small capacitance. A current source cannot be matched; therefore, the IF load must be selected according to some criterion other than conjugate matching. The only real option is to select a practically realizable load resistance that provides decent conversion efficiency and good stability. For small-signal FETs, this resistance usually is in the range of 50 to 100Ω. Lower impedances usually do not provide adequate gain, and higher impedances may introduce instability. Even though this is a mixer, it is still an active circuit, and active circuits can oscillate!

IF Gate Termination

One of the best things you can do for your active mixer is to short circuit the FET's gate at the IF frequency. The short-circuit gate termination prevents *amplifier-mode* operation at this frequency. Remember, the FET's transconductance waveform has a dc component, G_{m0} in (6.2), which results in an output term at the RF input frequency, $G_{m0}\, V_s \cos(\omega_s t)$. This is amplification, pure and simple. Amplifier mode gain can occur at the IF as well as the RF. The IF usually is a low frequency, so the FET may have quite a lot of IF gain. Any IF-frequency noise that leaks into the FET's input, often via the bias circuit, will be amplified and may degrade the FET's noise figure. Similarly, IF feedback may introduce instability. Short circuiting the gate at the IF frequency prevents this.

6.1.2 Conversion Efficiency

It is possible to develop an approximate expression for conversion gain. See [1] for the derivation. The conversion loss is

$$G_c = \frac{G_{m,\,max}^2 R_L}{16\omega_s^2 C_{gs}^2 (R_s + R_g + R_i)} \tag{6.5}$$

where $G_{m,max}$ is the peak transconductance, R_L is the IF load resistance, and the other parameters are elements of the FET's equivalent circuit, defined in Figure 2.18. This expression is based on the assumption that the gate is conjugate matched at the RF and biased at pinch-off.

6.1.3 Single-Device Equivalent Circuit

Many types of balanced FET mixers can be decomposed into single-device equivalent circuits. This representation has two advantages: first, it simplifies the design process, and second, it shows that balanced structures are really no different, in principle, from unbalanced ones. We will show how the decomposition process works when we examine the individual balanced mixers.

6.1.4 Other Configurations

Drain-Pumped Mixer

The urge to apply the LO to the FET's drain is almost irresistible. This mode of operation has an apparent advantage: the RF and LO are isolated. Unfortunately, drain pumping has a significant disadvantage: unless the FET is driven into its linear region over part of the LO cycle, the transconductance does not vary, and no transconductance mixing is possible. In a drain-pumped mixer (sometimes simply called a *drain mixer*), frequency conversion occurs because of a combination of resistive mixing, caused by the time-varying drain-to-source resistance, and transconductance mixing, caused by the decrease in transconductance when the FET drops out of saturation.

To put it simply, drain-pumped FET mixers don't work very well. They have never exhibited noise figures and gain as good as gate-pumped transconductance mixers, and their distortion is about the same. They do not have enough transconductance variation for efficient transconductance mixing, and the finite average drain-to-source resistance dissipates output power. Drain-pumped mixers would have withered from the technological consciousness a long time ago, but for a few early papers indicating that such mixers had unusually low IMD. Although these papers were not really conclusive, the idea that drain-pumped mixers have low distortion somehow became conventional wisdom. If you have heard this, forget it now, and the rest of your life will be much happier.

For more information on drain-pumped mixers, see [2].

Gate Resistive Mixer

Another possibility is to use the gate of the FET as a resistive mixer and the rest of the FET as an IF amplifier. This is another bad idea that didn't last very long. Aside from performance (which is not particularly good), the gates of modern MESFETs are not designed for this type of operation, and attempting it may damage the device.

If you are really interested in this mode of operation, in spite of this warning, see [3].

Source-Pumped Mixer

A third possibility is to apply the LO to the FET's source. Until we think a little less superficially about this configuration, it sounds like a distinct mode of operation. A little thought, however, reveals that a FET has three terminals, and any single excitation voltage can be applied only between a pair of them. In the gate-pumped mixer, the LO is applied between the gate and the source, and in the drain-pumped mixer, the LO is applied between the drain and the source. In the source-pumped mixer, the LO must be applied between the gate and the source or between the drain and the source.[2] In effect, it usually is applied between the gate and the source, and then the source-pumped mixer is equivalent to the gate-pumped mixer. The only differences are introduced by practical aspects of the design, such as the ease or difficulty of bypassing the source at the RF and IF frequencies.

These practical considerations sometimes make a source-pumped mixer worthwhile. For better or worse, source-pumped mixers are used only rarely.

6.2 SINGLE-FET MICROWAVE MIXER

6.2.1 Characteristics

Like single-diode mixers, a single-FET mixer exhibits high conversion efficiency and low noise at the expense of the benefits of balanced structures: even-order IMD rejection, LO AM noise rejection, and convenient separation of the LO, RF, and sometimes IF ports. One difficulty in single-device mixers, which can be avoided in balanced mixers, is the need for special structures to combine the RF and LO. Usually, some type of diplexer is used, although a directional coupler also can be used at the expense of high loss for either the RF or LO.

A second concern is LO leakage into the IF. The FET is, after all, an amplifying device. In diode mixers, the diodes absorb a large fraction of the LO power, but in FET mixers, the FET does the opposite: it amplifies any signal applied to the gate. Ideally, the circuit (usually a microstrip stub or a filter) that shorts the drain at the LO frequency should prevent LO-to-IF leakage, but microstrip circuit elements have

2. A third possibility is, of course, the gate and drain. We won't dignify this idea with a discussion.

relatively low Qs, and leakage occurs. Invariably, good LO-to-IF isolation requires some type of filter. In balanced mixers where the LO is applied out-of-phase at the gates, the LO-to-IF isolation can be very good. Avoiding LO-to-IF leakage is one especially good reason for using balanced mixers.

FET mixers are capable of conversion gain. When designed according to the criteria of Section 6.1, they also have good noise figures, and distortion also will be moderate. Typical performance for an X-band mixer using an ordinary small-signal FET and 6-dBm LO power is a conversion gain of 6 dB, noise figure of 4 dB, and output third-order intercept point of 12 dBm. High-performance HEMTs can provide conversion gain and low noise well into the millimeter-wave range.

The circuit described here is suitable for downconverter applications throughout the microwave and millimeter-wave range where the bandwidth is moderate and the IF frequency is low, no more than 10% of the RF. Because of the wide variety of possible approaches, we have not included a design for a combiner or an IF filter. In fact, it is best to design the mixer without these components and to add them as the last step of the design. For filter-design information, see Reference [4].

6.2.2 Description

The mixer circuit is shown in Figure 6.3. It is appropriate for use as a downconverter throughout the RF, microwave, and millimeter-wave ranges, as long as the bandwidth is moderate (~20% or less). For further information about the operation and design of these circuits, see [5].

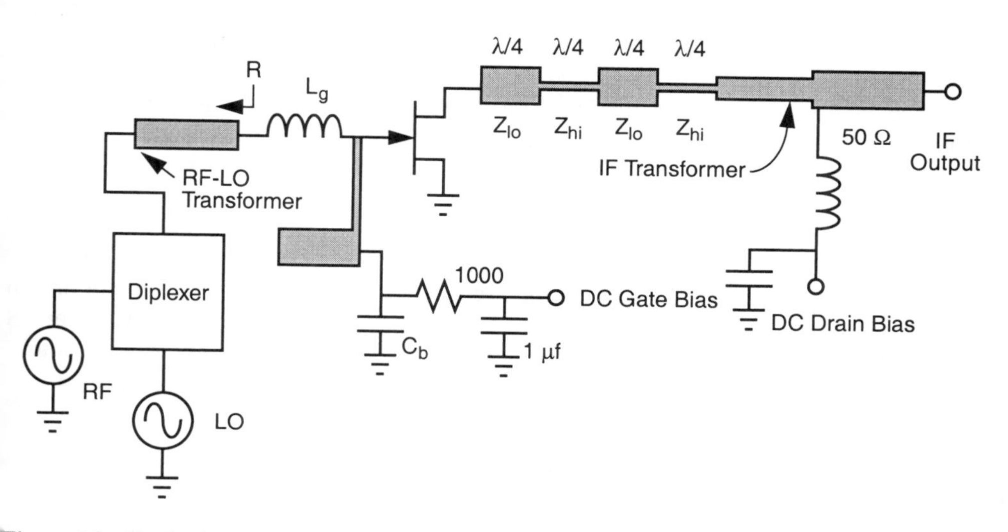

Figure 6.3 Single-device FET mixer. The $\lambda/4$ lengths are determined at the center of the RF-LO band, and C_b is the IF bypass capacitor. This circuit is appropriate for frequencies from the lower microwave region to millimeter waves.

The mixer consists of an input matching circuit, an output filter/matching circuit, bias circuits, and structures for short circuiting unwanted mixing products at the gate and drain. The structures in this design are ones we have found useful; however, anyone who understands their purposes, as outlined in Section 6.1, should have no trouble substituting other structures when appropriate.

We describe the design of the mixer only; the designer will have to come up with his own RF-LO diplexer. We'll add a few words of on the subject of LO injection at the end of Section 6.2.3.

This circuit is realized in microstrip. Microstrip (or some other planar transmission medium) is the only really practical way to realize such circuits.

6.2.3 Design

As with many microwave components, the design process largely is one of selecting appropriate source and load terminations for the input and the output. Additionally, we must provide optimum source and load terminations, as described in Section 6.1.1, at unwanted mixing frequencies and LO harmonics.

Device Selection

Conventional small-signal amplifier MESFETs are ideal for active FET mixers at frequencies up to about 20 GHz. There is little advantage in paying a premium for very low-noise devices; low noise figure in amplifier applications does not imply similarly low noise in mixers. At higher frequencies, HEMTs are best; at RF and lower microwave frequencies, MOSFETs may be appropriate. Power devices are sometimes used at low frequencies to achieve lower distortion, at the price of greater LO power requirements.

A device having a low (that is, more negative) pinch-off voltage requires more LO power but has lower distortion at its maximum LO level. HEMTs generally have greater distortion than MESFETs but require significantly less LO power.

Input Impedance

Because the gate circuit of a FET is not strongly nonlinear (as long as it is not driven into conduction), the input impedance of a pumped FET can be estimated from S parameters. It simply is the gate-input impedance of the device, biased near pinch-off, with a short-circuit on the drain. In terms of reflection coefficient, Γ_{in}, this is

$$\Gamma_{in} = S_{1,1} - \frac{S_{2,1} S_{1,2}}{1 + S_{2,2}} \tag{6.6}$$

The S parameters measured at low-noise bias, which is usually close to pinch-off, are adequate for this estimate.

A more accurate estimate can be found from nonlinear analysis of a simplified circuit. It is not necessary to create initial designs of the input or output matching circuits; just bias the device and apply RF and LO signals. The drain-to-source voltage should be set to the usual value for an amplifier (about 2.5–3.5V for most MESFETs), and the gate-to-source voltage should be set to pinch-off. Adjust the LO level until it is just short of drawing gate current. Use a large capacitor to short circuit the drain to the source or just connect the drain dc source directly to the FET's drain terminal. You should not observe much difference between the small-signal RF and large-signal LO input impedances at any frequency.

Whichever method you use, calculate the small- and large-signal impedances over the combined LO and RF passbands.

Output Impedance

The output impedance of a strongly pumped FET mixer is nearly infinite; there may be a small drain-to-source capacitance, 0.10–0.15 pF for small-signal FETs; unless the IF frequency is unusually high, this is negligible. Thus, selecting an output impedance involves selecting a value of resistance.

Equation (6.5) can be useful for this. R_L should be selected to provide the desired gain, within, of course, constraints of realizability. Remember that, in many systems, the highest possible gain is not necessary or even desirable, because it may exacerbate distortion and introduce instability. In a mixer, a conversion gain of a few decibels usually is adequate.

Nonlinear analysis can be used to optimize the load impedance. Do not try to calculate the output impedance directly. Instead, use the circuit you created to determine the input impedance. Adjust the bias and LO level as described in Section 6.1 and Figure 6.2. Connect a capacitor between the drain and source that is large enough to short circuit the RF and the LO but not the IF. If the IF frequency is too high to do this with a simple capacitor, just lower the IF frequency; the output is largely resistive at the IF, so it won't affect the results. Adjust the load resistance until the desired gain is achieved.

Input Matching Circuit

It is usually impossible to achieve a good input match over the entire LO band. Instead, use an inductor to resonate the input capacitance and adjust the real part of the source impedance to achieve adequate input bandwidth.

1. Calculate the required loaded input Q:

$$Q_{in} = \frac{f_0}{\Delta f} \tag{6.7}$$

where f_0 is the center of the combined RF-LO band and Δf is the bandwidth. This value of Q_{in} results in approximately a 3-dB roll-off at the band edges, which may be too much. You may wish to reduce Q_{in} accordingly.

2. Select the gate inductance:

$$L_g = \frac{-X_{in}}{2\pi f_0} \tag{6.8}$$

where $X_{in} = \text{Im}\{Z_{in}\}$, which normally is negative. Occasionally X_{in} will be positive at high frequencies, especially for packaged devices. In this case another matching approach should be used. Realize this inductance as a lumped element or as a distributed element, as appropriate for the operating frequency. See Sections 1.4 and 1.5.

3. Select the excitation source resistance, R:

$$R = \frac{-X_{in}}{Q_{in}} - R_{in} \tag{6.9}$$

where $R_{in} = \text{Re}\{Z_{in}\}$. At high frequencies, use a quarter-wave transformer to realize R. At low frequencies, use a wire-wound transformer.

4. Optimize this on the computer. Generate a simple series RLC model of the input from the calculated input impedance. You are attempting not simply to match the input, but to achieve uniform voltage across the capacitance (which represents the FET's gate-to-source capacitance) over the entire frequency range. This requires matching the input better at the high end of the band; some experimentation will be necessary.

Gate Bias Circuit and IF Short-Circuit Structure

The gate-bias circuit consists of a high-impedance series line and a low-impedance stub. Both are one-quarter wavelength long at the center of the RF-LO band. The series line's impedance should be as high as possible, and the stub should be as low as possible, within constraints of size and technology. This structure is good for about 40% bandwidth.

A chip capacitor is connected between the stub and ground (C_b in Figure 6.3). It should be large enough to short circuit the gate over the entire IF band. Ideally, the capacitor's series-resonant frequency (including the effects of the series line) should span the IF.

A resistor (on the order of 1,000Ω) and large-value capacitor (a few microfarads) complete the bias circuit. These provide additional filtering and reduce the danger of gate damage from electrical transients and static-electric discharge.

Output Filter/Matching Circuit

At the output we want to provide an appropriate load impedance, R_L, to the drain at the IF and to short circuit everything else, especially the LO and its harmonics. Figure 6.3 shows a simple output circuit that is quite effective. It consists of a cascade of low- and high-impedance microstrip sections, each one-quarter wavelength long at the center of the RF-LO band. This shorts the FET's drain effectively over a broad bandwidth. The lengths of the sections often can be adjusted to short circuit the second harmonic as well as the first. The third harmonic can be shorted with a stub, if deemed necessary. In FETs, the third harmonic is pretty weak, so shorting the first and the second usually is adequate.

Initially, the transmission-line sections' characteristic impedances should be chosen so that

$$Z_{0,l} = \sqrt{Z_{hi} Z_{lo}} \qquad (6.10)$$

where Z_{hi} and Z_{lo} are the characteristic impedances of the high- and low-impedance lines, respectively. Make these as high and as low as possible, respectively, within the constraint imposed by (6.10). At the IF frequency, where the sections are much shorter that one-quarter wavelength, the structure will behave approximately as a line of characteristic impedance $Z_{0,l}$.

If $R_L = Z_{0,l} = Z_0$, where Z_0 is the IF port impedance (usually 50Ω), this completes the initial design. If not, $Z_{0,l}$ can be made equal to the impedance of a quarter-wave IF transformer used to achieve the desired value of R_L. If the structure is less than one-quarter wavelength long at the IF frequency, the rest of the transformer is located on the output side of the structure. In this case, (6.10) becomes

$$R_L Z_0 = Z_{hi} Z_{lo} \qquad (6.11)$$

These steps result in a crude initial design. The circuit should be computer optimized to achieve a good short circuit over the RF-LO band and at least its second harmonic and to present the appropriate load resistance R_L to the FET at the IF.

The drain bias circuit is simply an RF choke and bypass capacitor. These usually can be lumped elements.

Nonlinear Analysis of the Mixer

Nonlinear analysis is very effective for optimizing the design of active mixers. It is essential, however, to optimize the individual parts of the mixer—the input, output, and bias circuits—before performing a nonlinear analysis.

Active mixers do not demand highly accurate device models; any model that is reasonably representative of the device is adequate. The important elements of the equivalent circuit, for accurate results, are the gate, source, and intrinsic resistances and the gate-to-source capacitance. It is essential that the model reflect the peak transconductance accurately and that the transconductance be zero at pinch-off. Be careful: although this requirement seems obvious, many parameter-extraction methods do not result in a model that accurately reproduces these parameters.

Diplexer

This mixer requires some type of circuit to combine the RF and LO signals. The most common tool for doing this is a filter diplexer, which consists of two filters, one for the RF and one for the LO, whose passbands do not overlap. The outputs are connected in parallel, so the output impedance of one filter becomes part of the other, and the design of both filters must account for this.

Filter diplexers can be tricky to design if the RF and LO bands are closely spaced in frequency. Achieving adequate RF rejection in the LO filter and LO rejection in the RF filter requires complex structures, which usually cause the diplexer to be much larger and more complex than the rest of the mixer. Chapter 16 of Matthaei, Young, and Jones [4] has everything you need to know about diplexers for mixer design.

Another approach is to use a directional coupler. Generally, the RF is applied to the "through" port and the LO to the coupled port. This is much simpler than a diplexer but results in several decibels of LO power loss and not insignificant RF loss. For example, if 1 dB of RF loss is acceptable, the coupling—and therefore the LO loss—can be no stronger than –7 dB. That much LO loss is usually unacceptable.

A third possibility is to apply the LO to a terminal of the mixer other than the gate. In Section 6.1 we cruelly disparaged the practice of applying the LO to the drain and weren't terribly enthusiastic about applying it to the source. Even though source pumping is done from time to time, it really is not as promising as it might appear. First, for good performance, we must bypass the source (short circuited, to be precise) at the RF frequency. If the RF and LO are close in frequency, this structure may be as complex as the LO filter in a diplexer. Second, for less obvious reasons, the gate must be bypassed (again, short circuited) at the LO frequency, so the LO effectively is applied between the source and the gate. As with a diplexer, we still require two filters; we just have relocated them.

A fourth possibility is to use a different type of mixer. A dual-gate device may be a good solution to this problem, as might a balanced mixer. We examine these options in Section 6.3.

6.2.4 Variations

The output circuit shown in Figure 6.3 is rather large, so it is tempting to use a simple quarter-wavelength open stub to provide the LO short circuit to the drain. A stub requires less space than the more complex IF filter in Figure 6.3, but it introduces a few new problems. First, the Q of a microstrip stub is not great, so LO rejection is at best modest. Second, a quarter-wavelength open-circuit stub, which shorts the LO at its fundamental frequency, does not short circuit the drain at the second harmonic, so substantial second-harmonic leakage may occur. This can cause spurious responses and spurious signals in the IF amplifier or even saturate it.

At millimeter wavelengths, HEMTs make very good active mixers. At lower frequencies MOSFETs and JFETs can be used. At frequencies below about 2 GHz, microstrip structures may become impractically large. In this case a wirewound transformer can be used for the LO-RF transformer and a lumped parallel L-C circuit, tuned to the IF, may be adequate to short-circuit the drain at the LO and RF frequencies. RF transformer design is almost a specialty in itself; for an introduction to the subject, along with all the classical references, see Reference [6].

As with an amplifier, this circuit can be self biased. This requires a resistor in series with the source and a bypass capacitor in parallel with the resistor. The value of resistance R_b is approximately

$$R_b \approx \frac{0.33\, I_{dss}}{V_p} \tag{6.12}$$

where we have estimated the drain current of the pumped FET as approximately one-third of the FET's I_{dss}. The capacitance's reactance should be much smaller than the resistance at the lowest IF frequency, and its self-resonant frequency must be high enough so that the capacitor's series inductance remains insignificant.

Bipolar devices—BJTs and HBTs—also can be used for active mixers. Bipolar devices rarely are used for single-device mixers, probably because the Gilbert cell (Section 6.7) is so much better. Because the low-frequency gain of a bipolar device is very high, amplifier-mode problems (Section 6.1.1) at the IF may be more severe. The Gilbert cell reduces these problems significantly.

6.2.5 Cautions

Don't attempt to operate an active mixer at a low LO level to reduce its gain. If LO power is expensive, use a smaller device or one having a lower pinch-off voltage and pump it to its limit. HEMTs require very little LO voltage at their gates to achieve maximum transconductance variation, so they are ideal for use when little LO power is available. Adjust R_L to achieve the desired gain; this provides the best gain stability.

Do not drive the gate too hard with the LO. Monitor the gate current. If the FET starts to draw gate current, the reliability will suffer, as will its noise figure and intermodulation intercept points. The resistor in the dc bias circuit, shown in Figure 6.3, will help prevent damage from overdrive. Self-bias also will help prevent such damage.

6.3 SINGLE-DEVICE, DUAL-GATE MIXER

6.3.1 Characteristics

Silicon MOSFET dual-gate mixers have been common in mobile and hand-held radio transceivers since the 1960s. They exhibit decent noise figures and very low distortion, and, most important, they provide good LO-to-RF isolation without the use of filters. Microwave dual-gate MESFET mixers are newer, dating from the mid-1980s, and exhibit much the same characteristics. Their conversion gain and noise are somewhat worse than single-gate MESFET mixers, 1 to 2 dB at best, but the output intermodulation intercept point often is a few decibels higher.

A dual-gate mixer is not really a transconductance mixer. It is more like a drain mixer and has many of the same characteristics; this explains the lower performance. Still, with careful design, the performance can be adequate for many purposes, and the lack of baluns and an LO-RF diplexer allows a compact circuit, ideal for microwave monolithic integrated circuits (MMICs).

6.3.2 Description

A dual-gate device can be modeled as two cascode-connected single-gate devices. This structure, shown in Figure 6.4, includes matching circuits at both gates and the drain. It also has bypass filters at the drain and the second gate. By convention, gate 1 is the gate closest to the source and gate 2 is the gate closest to the drain.

The lower FET in the cascode, FET 1, provides the mixing. It essentially is a drain-pumped mixer, with the LO provided by the upper FET, FET 2, which operates as a source follower. The two FETs operate as a cascode amplifier at the IF. This

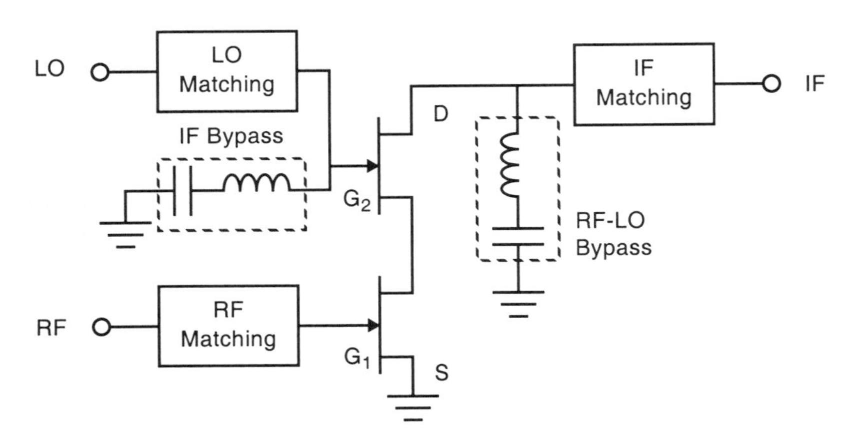

Figure 6.4 The dual-gate FET mixer can be modeled as two cascode-connected single-gate FETs. The drain LO-RF bypass serves the same purpose as the output filter in the singe-gate FET mixer; the IF bypass allows the FETs to operate as a cascode amplifier at the IF.

requires that the gate of FET 2 be grounded at the IF frequency; that ground connection is provided by the IF bypass.

The drain filter serves the same purpose as the one in the single-gate FET mixer: it short circuits the RF and LO at the drain, keeping the large-signal drain voltage constant, at the dc-bias value, over the entire LO cycle. This maximizes LO-to-IF isolation and prevents small-signal feedback effects.

Figure 6.5 shows the bias and LO operating regions of the individual FETs. The gate bias of FET 1 usually is fairly high, about 0V for a MESFET or JFET and several volts positive for a MOSFET. FET 1 therefore drops into its linear region when its drain voltage (the voltage at the floating node connecting the two devices) is low and becomes current saturated when its drain voltage is high. The gate bias of FET 2, on the other hand, is adjusted so that the device switches on and off every half LO cycle. This creates a large LO voltage at the drain of FET 1, resulting in efficient mixing (or at least as efficient as possible). Note that the gate-to-source voltage of FET 2 depends on the drain voltage of FET 1, which is a floating quantity:

$$V_{gs2} = V_{g2} - V_{ds1} \tag{6.13}$$

where V_{gs2} is the gate-to-source voltage of FET 2, V_{ds1} is the drain-to-source voltage of FET 1, and V_{g2} is the voltage applied to the gate of FET 2.

The drain-to-source conductance of FET 1, the mixing device, is not low over the whole LO cycle. Over approximately half the cycle, when the FET is in its linear

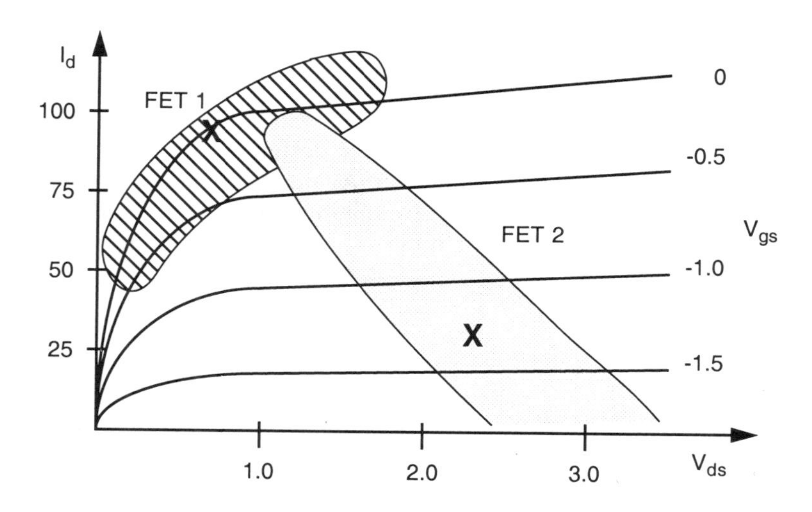

Figure 6.5 Drain I/V characteristic of one of the cascode-connected MESFETs in the dual-gate device and the FETs' LO operating regions. The 'X' shows the approximate bias points for each device. The cross-hatched area shows the operating region for the lower FET (FET 1) and the shaded region for the upper FET (FET 2).

region, the conductance rises to a high value. The result is a low average drain-to-source resistance that effectively is in parallel with the device. The loss in this resistance, which does not exist in the single-gate mixer, is one reason for the poorer noise figure and conversion gain of the dual-gate mixer.

It is possible to develop a simple, approximate expression similar to Equation (6.5) for the conversion gain of a dual-gate FET mixer:

$$G_c = \frac{G_{m1}^2 \dfrac{G_{m2}^2}{(G_{o1} + G_{m2})^2} R_L}{4\omega_s^2 C_{gs1}^2 (R_g + R_i + R_s)\left(1 + \dfrac{C_{gs2}^2 \omega_{IF}^2}{G_{m2}^2}\right)} \tag{6.14}$$

where G_{m1} is the *fundamental-frequency* component of the transconductance of FET 1, G_{m2} is the *average* transconductance of FET 2, G_{o1} is the average drain-to-source conductance of FET 1, C_{gs1} is the gate-to-source capacitance of FET 1, C_{gs2} is the gate-to-source capacitance of FET 2, and the resistances are those of FET 1. As before, ω_s is the RF frequency and ω_{IF} is the IF.

Unfortunately, some of the parameters of (6.14) are difficult to estimate accurately. G_{m1} is not 0.5 $G_{m,max}$, as was estimated in (6.5); 0.3 $G_{m,max}$ is a closer estimate. A decent estimate for G_{o1} is $0.5 / R_{ds}$, where R_{ds} is the drain-to-source resistance of FET 1 in its linear region at minimum drain voltage, typically 100Ω. It is reasonable to assume that $C_{gs2} = C_{gs1}$. Note that, with $G_{m1} = 0.5\ G_{m,max}$, $G_{o1} = 0$, and $\omega_{IF} \to 0$, (6.14) reduces to (6.5).

6.3.3 Design

In spite of its large and obvious differences, the design of this mixer is surprisingly similar to the design of the single-gate mixer. Again, we must determine the RF and LO input impedances and the IF output terminating impedance. We also must design matching circuits and provide short-circuit terminations to certain points at certain frequencies.

Device Selection

As with single-gate devices, small-signal MESFETs are most appropriate. You will have a hard time finding dual-gate HEMTs, although such devices may be available in ICs. Of course, dual-gate MOSFETs are ideal for applications at frequencies of a few hundred megahertz and below.

RF Input Impedance

The RF input impedance of FET 1 is dominated by the FET's resistive parasitics and gate-to-source capacitance, and the estimate of input impedance for the single-gate device (see Section 6.2.3) is applicable here. This impedance can be found from S parameters, as in (6.6), or from a small-signal equivalent circuit of the device, where

$$Z_{in} = R_s + R_i + R_g + \frac{1}{\omega_s C_{gs}} \tag{6.15}$$

and ω_s is the RF frequency. C_{gs} is evaluated at the gate-bias point. This expression represents the FET alone; if a packaged device is used, package parasitics must be included as well.

LO Input Impedance

FET 2 is an LO source follower, and like all source-follower circuits, its input impedance is very high. It essentially is an open circuit with a small shunt capacitance, usually about 0.1 pF for ordinary small-signal MESFETs. The value of the capacitive reactance is difficult to predict accurately by approximate methods. Nonlinear circuit analysis is necessary.

IF Output Impedance

Like the single-gate device, the IF output impedance of the dual-gate device is an open circuit in parallel with a small capacitance, usually less than 0.1 pF for small-signal MESFETs. This impedance clearly is impossible to match. Instead of attempting a conjugate match, the goal of IF design is to select an appropriate load impedance. The method for doing this is not clear, however; (6.14) may be helpful, but don't be surprised if it is not very accurate. Often, the most practical value is simply the 50Ω port impedance.

Unless the IF frequency is very low, the capacitance of the stub used to short circuit the FET's drain at the LO frequency may be a large part of the IF output impedance. Be sure to account for this in designing the IF.

RF Matching Circuit

The design approach for the input matching circuit in Section 6.2.3 is applicable here. The design for this mixer is somewhat easier, since the matching circuit need not cover the LO frequency band. At low frequencies C_{gs1} may be negligible and a matching circuit may not even be necessary.

LO Matching Circuit

The LO input impedance also is very high and impossible to match. Instead, the goal of the LO design should be to achieve the highest possible source impedance driving the gate of FET 2. This impedance is limited primarily by the input capacitance, C_{in}, (which usually is somewhat lower than C_{gs2}). The source impedance can be increased, by means of an appropriate transformer, as long as the LO source resistance is much less than $1/\omega_{LO} C_{in}$. If an inductor is used to resonate C_{in}, the source resistance can be higher, but still must be low enough to provide adequate bandwidth.

It is virtually impossible to match the LO port, and the LO source normally must be designed to tolerate a high VSWR. If the high VSWR is not tolerable, the only practical solution, other than an isolator, is to connect a resistor between the gate and ground, with appropriate dc blocking, and to match the LO to that resistance. The higher the resistance, the better; the input capacitance and the required bandwidth will limit the value, as will practical constraints of realizability. *Warning*: the use of a loading resistor effectively reduces the LO power by 6 dB, compared to a port having no resistor. Remember, loading the output port of any source with a matched resistor reduces the voltage by one-half, which is equivalent to a 6-dB loss.

IF Circuit Design

The IF output effectively is a current source in shunt with a small capacitance. Usually, the mixer's IF load impedance is simply the 50Ω port impedance. If the mixer is used in an MMIC that includes an IF amplifier, other load impedances may be practical. The input impedance of a low-frequency FET IF amplifier usually is quite high, so a resistor can be connected in shunt with the IF amplifier input to set the amplifier's input impedance and thus the mixer's load impedance. This resistance should be relatively high; the higher its value, the less effect on the circuit's noise figure and the greater the gain. The shunt capacitance of the IF and mixer output limit the value that can be used.

Remember, the IF-frequency reactance of the stub used to short circuit the dual-gate FET's drain must be included in the mixer's output capacitance.

Bypass Filters

The drain must be shorted at the fundamental LO frequency and preferably two or three harmonics. The output matching filter design from Section 6.2.3 can be used, but it may be too large for an IC. If available space is not great, a stub can be used; the disadvantages of a stub are the same as those for a single-FET mixer. In a balanced circuit, this short circuit may be inherent.

The second gate must be shorted at the IF frequency. If the IF frequency is relatively low, a simple inductor may be adequate. Otherwise, appropriate structures are either a lumped series-resonant circuit or a shorted stub one-quarter wavelength long at the LO frequency.

Analysis

This is the time to think seriously about buying that harmonic-balance simulator you've always wanted. Dual-gate FET mixers are fairly complex circuits, and simplified analysis, at best, is only marginally adequate. Nonlinear analysis by computer really is necessary.

6.3.4 Design Example

This circuit is complex, and the description has been very abstract. Perhaps an example will make everything a bit more concrete.

We design a mixer with moderate but overlapping RF and LO bandwidths, both 9 to 11 GHz, and an IF from 0.1 to 1 GHz. Since the RF and LO bands are identical, an unbalanced single-gate mixer is not practical. An ordinary 300-μm dual-gate FET is available. Our approach is to begin with an approximate design and to fine-tune it using nonlinear analysis. Of course, the latter stage requires a model of the device and a lot of expensive software. The alternative is to do the approximate design, fabricate the circuit, and to optimize it manually.

Figure 6.6 shows the basic circuit.

Initial Investigation

The device has the following parameters:

- $G_{m,max} = 0.07$ S
- $R_{ds} = 100\Omega$

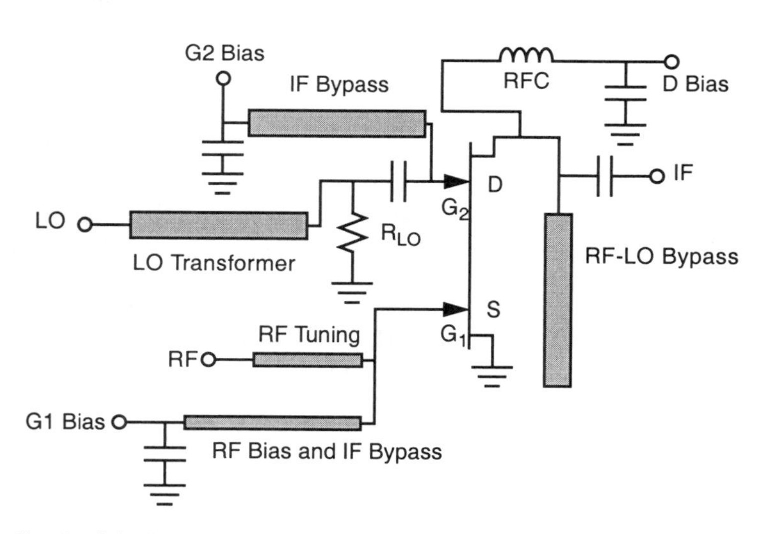

Figure 6.6 Circuit of the dual-gate MESFET mixer described in Section 6.3.4.

- $R_s = 1\Omega$
- $R_i = 4\Omega$
- $R_g = 2\Omega$
- $C_{gs} = 0.30$ pF
- $V_p = -2.0$V
- $R_L = 50\Omega$

These parameters apply to either FET in the cascode model of the dual-gate device. From (6.14), we find that the device is capable of 1.5-dB conversion gain at the center of the RF/LO frequency range and the top of the IF range. Remember, (6.14) assumes that the RF port is matched, which probably will not be possible. To determine the conversion gain when the input is mismatched, simply estimate the RF input reflection coefficient, Γ, and subtract the reflection loss in decibels, $20 \log(1-|\Gamma|^2)$. With a dB or two of reflection loss, which we'll estimate more accurately later, we expect approximately unity gain. This is adequate for many applications.

The pinch-off voltage is -2.0V, so we need a peak LO gate voltage of 2V to turn the device on completely. The open-circuit voltage V_s of a source is

$$V_s = \sqrt{8RP_{av}} \tag{6.16}$$

where R is the source resistance and P_{av} is its available power. From (6.16) we find that the 2V open-circuit voltage, from a 50Ω source, requires 10 dBm of LO power. If a 50Ω resistor were connected to the gate to minimize the LO VSWR, 16 dBm would be required. This is a little high. To reduce the LO power requirement, while maintaining a good LO VSWR, we should increase R and use a transformer to match it to our 50Ω source or avoid using a loading resistor entirely.

RF Design

The RF circuit is designed in a manner similar to that of the single-gate mixer (Section 6.2.3). We use a series input inductor to resonate C_{gs}, and we select the real part of the RF source impedance to provide a Q low enough to cover the entire band. To keep the circuit small and simple, we want to drive the gate directly from our 50Ω source, without any transformation.

With 0.3 pF C_{gs}, the input reactance is –53Ω at bandcenter. Thus, we need an inductor whose reactance is 53Ω, 0.85 nH, to resonate it. The total resistance in the loop is $R + R_s + R_i + R_g$, 57Ω. Thus, the Q is about 1.0, providing more than adequate bandwidth. When the capacitance is resonated, the input impedance is real, and the input VSWR is $R / R_s + R_i + R_g$, about 7.1, a reflection loss of 3.6 dB. That leaves us with –2.1 dB conversion loss (1.5-dB matched conversion gain minus 3.6-dB input reflection loss). If higher gain were desired, an input RF transformer could be used to reduce the source impedance. This would increase the input Q but, because the bandwidth is greater than necessary, there is room to do it.

Because of the high frequency, we realize the series inductor by a transmission line. The impedance of the line is approximately $Z_0 \tan \theta$, where θ is the electrical length and Z_0 is the characteristic impedance. Z_0 should be as high as possible. If $Z_0 = 100\Omega$, we need $\theta = 28$ degrees to achieve our 53Ω reactance.

The rest of the circuit is for dc bias and an IF bypass on the gate. It consists of a high-impedance RF bias line one-quarter wavelength long at the center of the RF band and a bypass capacitor. To short circuit the gate effectively at the IF frequency, the bypass capacitor must have a low reactance at the lowest IF frequency.

LO Design

First, we need to select the value of the loading resistor. The higher its value, the lower the LO power; however, this resistor, in combination with the input capacitance, creates a real pole that limits bandwidth. We need

$$\omega_{RF} \ll \frac{2}{R_{LO} C_{in}} \tag{6.17}$$

where R_{LO} is the value of the loading resistor and C_{in} is the input capacitance. (The resistance determining the bandwidth is actually $R_{LO} / 2$, because the transformed source impedance is in parallel with R_{LO}; hence the factor of 2.) Finally, the transformer used to match the 50Ω source to this resistance also restricts bandwidth somewhat, and the chip resistor has parasitic inductance and capacitance that increases as the resistance increases and the resistor becomes physically larger. These factors limit the practical loading resistance to about 100Ω. Still, 100Ω is a significant improvement; (6.16) shows that, for a given source voltage, doubling the source resistance halves the required LO power.

Even if C_{in} is around 0.1 pF, the real pole created by a 100Ω R_{LO} is about 30 GHz, so this value of R_{LO} will not restrict bandwidth or reduce the LO voltage at the gate. The bandwidth of the matching transformer can be found from Table 6.02-2 in Matthaei, Young, and Jones [4]. The band-edge VSWR is 1.12, quite adequate. The transformer's characteristic impedance is simply $(Z_0 R_{LO})^{0.5}$, which is 70.7Ω, and it is one-quarter wavelength long at the center of the LO band.

The IF bypass is a quarter-wavelength stub. The characteristic impedance of this stub must be high enough so it does not restrict the LO bandwidth but low enough so its impedance at the upper end of the IF band is not too great. The length must be 90 degrees at 10 GHz, so it will be 9 degrees at 1 GHz, the top of the IF band, and 81 and 99 degrees long at the edges of the LO band. At the IF, the stub impedance is $Z_0 \tan(9)$ or $0.16\, Z_0$. At the LO band edges, it is $Z_0 \tan(81) = -Z_0 \tan(99) = 6.3\, Z_0$. If we select $Z_0 = 60\Omega$, we have less than 10Ω IF reactance but almost 400Ω in shunt with the gate at the LO band edge. This seems adequate. Z_0 can be optimized further when the circuit is simulated on the computer.

The remaining capacitors in the LO circuit are blocking and bypass capacitors.

IF Design

Again, for simplicity, we choose $R_L = 50\Omega$, the load impedance. The gain could be increased by increasing R_L to about 100Ω. Because of the wide 0.1 to 1 GHz bandwidth, a transmission-line transformer is impractical but a wire-wound transformer could be used [6].

The RF/LO bypass stub is one-quarter wavelength long at 10 GHz. The lower the characteristic impedance of this stub, the broader the bandwidth. If we make it 30 Ω, the impedance at the band edge, 30 cot(81), is only about 5Ω. That should be fine. A lower impedance probably is not practical, because the low-impedance stub becomes impractically wide. At the IF frequency, this stub is a capacitor. It's reactance at 1 GHz is 30 cot(9), or 190Ω. This is roughly equivalent to 0.8 pF and clearly is the dominant component of the IF capacitance. Still, it is a tolerable shunt reactance, causing only a 1.3 VSWR at 1 GHz if a 50Ω load is used.

The drain bias circuit uses an RF choke and capacitor, sized in the conventional manner.

Final Note

It is important to recognize that this design includes some pretty grand approximations. It should be viewed as an initial, "first cut" design, which must be modified substantially through nonlinear analysis, manual tuning, or both.

Tuning and Adjustment

In a complex circuit like this, it is important to decide at the outset what to tune and what to leave alone. All the parts of the circuit affect all the performance characteristics, so mindless tuning to improve one characteristic (for example, the gain) can hurt another (perhaps the noise figure or isolation). Part of the secret in making things like this work is to figure out ways to adjust one characteristic and to optimize it without making a mess of the rest. Here are some suggestions:

- Start with the bypass stubs. It should be possible to design these precisely. However, ground vias in short-circuit stubs and fringing fields at the end of open-circuit stubs can change their resonant frequency. Sweep the LO frequency across the band and adjust the IF stub for maximum LO rejection at 10 GHz and the IF bypass stub for minimum VSWR at 10 GHz. *After these adjustments have been made, don't touch the stubs.*
- Adjust the RF tuning line for best gain across the band. The minimum VSWR may not occur at band center; it probably will be closer to the high end of the band.
- Adjust the LO level dc bias to optimize the conversion gain; you may have to go back and tweak the RF tuning a little. If the bias voltages are far from the points indicated in Section 6.3.3 and Figure 6.5, something has gone wrong.

With these adjustments, the mixer should work well. Little effort should be needed to make it "play"; if a lot of effort is needed, or the mixer exhibits strange behavior, there probably is a design error. If so, go back and find the problem; don't just "tweak" mindlessly at the circuit and settle for the mediocre performance that inevitably will result.

6.3.5 Variations

At RF and lower microwave frequencies, MOSFET or JFET dual-gate mixers are practical. The design approach is essentially the same as for the MESFET mixer; the only difference is (1) enhancement-mode MOSFET devices require positive gate bias, and (2) at the lower frequencies lumped-element matching, bypass, and bias circuits should be practical.

6.3.6 Cautions

Again, we emphasize the point that this design procedure is very approximate. It really must be optimized by nonlinear analysis.

Don't expect a dual-gate FET mixer to work as well as an optimized single-gate mixer. Dual-gate mixers represent a clear trade-off of overall performance for LO-to-RF isolation and size.

The LO-to-RF isolation of the dual-gate mixer is its greatest advantage over other FET mixer circuits. This isolation is not perfect, however. Because of capacitance between the gates (which are parallel strips of metal) and coupling within the FET, the isolation is rarely better than 20 dB at 10 GHz. At higher frequencies, isolation may be even worse.

6.4 SINGLY BALANCED FET MIXERS

6.4.1 Characteristics

Singly balanced FET mixers have essentially the same benefits, compared to their unbalanced brethren, as diode mixers: rejection of AM noise on the LO, rejection of certain (but not all) even-order intermodulation products and spurious responses, inherent RF-to-LO isolation, and, when properly designed, inherent rejection of LO energy from the IF.

Balanced FET mixers can have conversion gain and noise figures almost identical to those of single-device mixers; the loss of passive baluns or hybrids may increase noise figures and reduce conversion gain a few tenths of a decibel. Intermodulation intercept points are nearly 3 dB greater, however, as are 1-dB compression points. This increase in large-signal handling has a price: LO power requirements also are 3 dB greater.

6.4.2 Description

Figure 6.7 shows a singly balanced FET mixer. It uses a 180-degree hybrid for the RF and LO; 90-degree hybrids also can be used, but 180-degree rat-race hybrids are by far the most common. See Section 3.3 for a discussion of the disadvantages of 90-degree-hybrid diode mixers; they apply to active mixers as well.

The balanced FET mixer consists of little more than two single-device mixers connected by a set of hybrids or, occasionally, baluns. The best way to design a balanced mixer is first to design a single-device mixer and connect two of them together. Anyone who understands the single-FET mixer also understands almost everything about the balanced mixer.

One immediate difference between singly balanced diode and FET mixers is that the FET mixer uses an IF hybrid. The diode mixer (see, for example, Figure 3.2) has no IF hybrid, but the two diodes' polarities are reversed. FETs cannot be reversed, so they must be oriented identically, necessitating the IF hybrid. Its hybrids represent the major disadvantage of this circuit; especially at low frequencies, the IF hybrid usually is the largest component in the circuit.

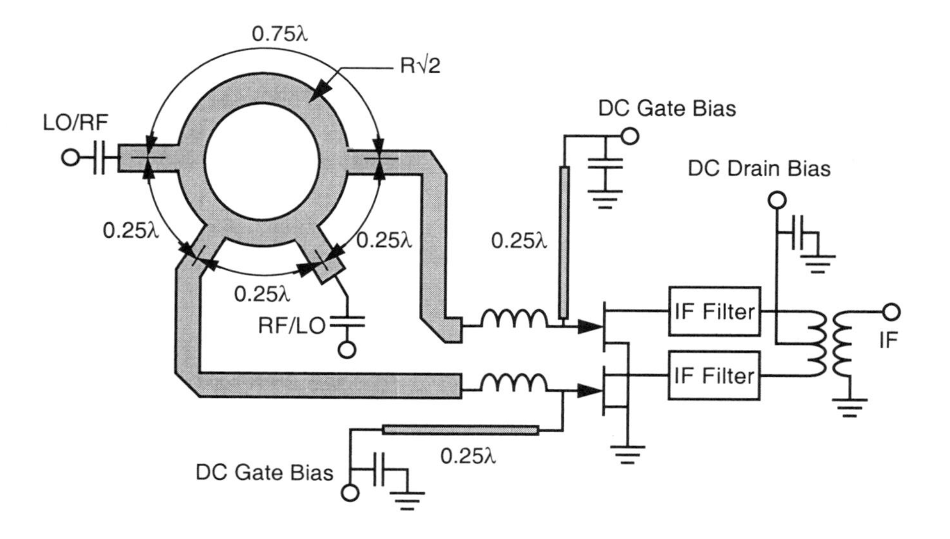

Figure 6.7 A singly balanced FET mixer consists of two single-FET mixers connected by a hybrid. The details of the individual single-FET mixers are not shown, since they are essentially the same as the circuits described in Section 6.2. The interconnections from the RF-LO hybrid to the FET gates can include transformers or other structures for impedance matching. Dimensions given as fractional wavelengths are evaluated at the center of the RF-LO band.

6.4.3 Design

Designing the balanced mixer involves designing the individual single-FET mixers and the set of hybrids to connect them. In Section 6.2.3 we discussed the design of the individual mixers, so Figure 6.7 does not show these mixers in detail.

RF/LO design

The input matching circuits of the individual mixers are designed as described in Section 6.2.3. The process involves using an inductor to resonate each FET's input capacitance and a transformer to adjust the Q of the input resonance so both the LO and RF bands are covered. Equations (6.7) through (6.9) are used to determine these quantities. There is no point in using a broader-bandwidth matching circuit, because the bandwidth is limited by the input hybrid.

The FETs' gates are connected to the hybrid by lengths of transmission line. The lines can be quarter-wave transformers that realize the RF-LO source impedance given by (6.9). The gate-bias stubs are used to short-circuit the IF at the FETs' gates; the bypass capacitor at the end of each stub must be large enough to short-circuit the lowest IF frequency.

It is necessary to provide dc blocks at the LO and RF inputs. In contrast to the rat-race diode mixer, there should be little IF energy on the hybrid, and IF blocks are not needed in the RF or LO ports.

IF Circuit

The options for the IF circuit given in Section 6.2.3 are valid here as well. Most important is the LO-frequency short circuit provided to the drains by the IF filter. It has a strong effect on all aspects of performance.

The IF currents from the two FETs differ in phase by 180 degrees. Combining them requires a balun or hybrid. The simplest IF combiner is a wire-wound transformer; the highest practical frequency for wire-wound transformers is about 2.0 GHz, although higher-frequency transformers occasionally are realized.

If the LO is applied in phase at the gates (that is, to the port in Figure 6.7 marked RF/LO), the LO leakage in the IF may cancel somewhat in the transformer. Because the LO probably is outside the transformer's passband, the cancellation may not be great. The poor coupling of the transformer at high frequencies alone may provide substantial LO rejection. A transformer also may increase the IF load impedance at the drains, which will increase gain.

At higher IF frequencies a conventional 180-degree microwave hybrid can be used. It is likely to be large. Lumped-element hybrids (described in [7]) have good bandwidth and are much smaller.

6.4.4 Variations

To minimize circuit size, balanced FET mixers sometimes use active baluns or hybrids. These often have substantial noise and limited dynamic range. As a result, they may increase the noise figure, distortion, and 1-dB compression points of the mixer. The differential circuit described in Section 6.5 may be a better choice than a singly balanced mixer using active baluns.

Often a differential IF amplifier follows a balanced FET mixer. That way, the IF amplifier operates as a combiner as well as an amplifier.

6.4.5 Cautions

Do not try to tune individual mixers of a balanced pair; it is almost impossible to tune the individual mixers without unbalancing the circuit. Tuning may improve the gain or noise figure a little, but it usually destroys the characteristics that attracted you to a balanced mixer in the first place. Clearly, the design of each single-FET module should be optimized before it is connected into a balanced circuit. If the hybrid is properly designed and the mixers are optimized, no tuning of the balanced circuit should be necessary. In fact, it is rarely necessary to simulate the entire balanced mixer on the computer; simulating the individual single-device mixer stages usually is adequate.

Because the FETs are not reversed, the spurious-response rejection properties of the singly balanced FET mixer are the opposite of the singly balanced diode mixer (Section 3.2.2). Specifically, applying the LO to the out-of-phase port (sometimes called the *delta* port) of the hybrid, marked LO/RF in Figure 6.7, rejects (1, 2) spurious responses but not (2, 1). If the LO and RF ports are reversed, the (2, 1) response is rejected but not (1, 2).

6.5 SINGLY BALANCED DIFFERENTIAL MIXER

6.5.1 Characteristics

This mixer provides essentially the same benefits as any other type of singly balanced mixer. Additionally, like most active FET mixers, it is capable of conversion gain. It can be realized in MOSFET technology at lower frequencies or in MESFET technology at microwave frequencies.

6.5.2 Description

This mixer, shown in Figure 6.8(a), looks a lot like a differential amplifier. The RF is applied directly to the lower FET, and the LO is applied through a balun or hybrid to the upper FETs.

The upper FETs of this mixer operate as a commutating analog switch. These FETs connect the drain of the lower FET alternately, every half LO cycle, to the

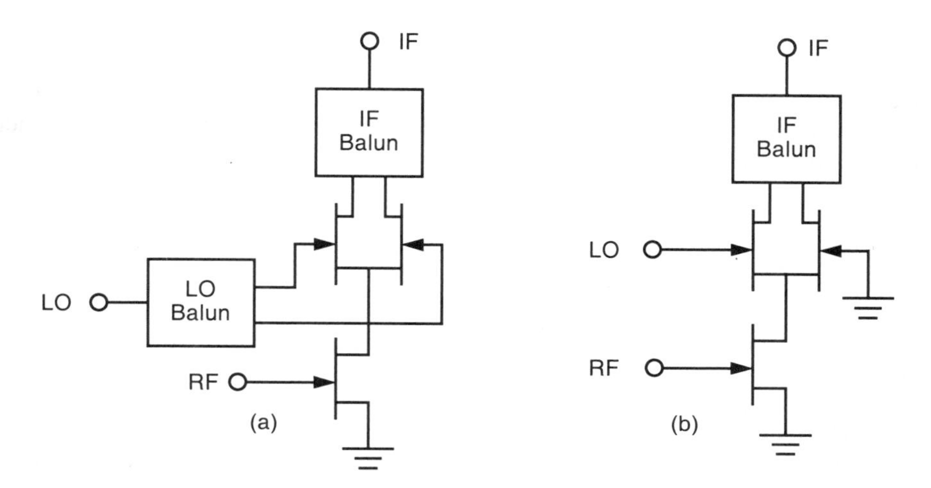

Figure 6.8 Singly balanced differential mixers. The upper FETs operate as commutating switches, while the lower ones operate simply as amplifiers. Circuit (a) is the classical configuration; (b) can be used with JFETs or MOSFETs. Because of a MESFET's low drain-to-source resistance, (b) usually is a poor choice for MESFET mixers. Matching and bias circuits are not shown.

inputs of the IF balun. The node connecting the sources of the upper FETs is a virtual ground for the LO; therefore, the LO simply switches the two upper FETs on and off on alternate half cycles.

It is tempting but incorrect to view this circuit as a variant of the dual-gate mixer. In the dual-gate mixer, the upper FET remains in current saturation over most of the LO cycle, and the lower FET is driven into its linear region over part of the LO cycle. Here, the upper pair of FETs may drop into its linear region over part (or even all) of the LO cycle, and the lower FET remains in saturation throughout. This mixer is most like an amplifier (the lower FET) in series with an LO-operated switch.

6.5.3 Design

The key to understanding this circuit is to recognize that there is no LO voltage at the node connecting the sources of the upper FETs to the drain of the lower FET.[3] Therefore, this point is a virtual ground for the LO. Recognizing this, we can simplify the LO circuit to a single FET driven by the LO source. Likewise, the RF equivalent circuit is just a single *unpumped* FET whose load is the impedance measured at the sources of the upper FETs. This is a very low impedance.

---

3. If you don't understand this, consult any basic electronics book on the subject of differential amplifiers.

LO Design

Because of the virtual ground at the drains, the LO input impedance is simply the impedance of the input loop of the FET. The LO input impedance of each device, Z_{LO}, is approximately

$$Z_{LO} \cong R_s + R_g + R_i + \frac{1}{j\omega_{LO} C_{gs}} \qquad (6.18)$$

where ω_{LO} is the LO frequency. A 180-degree hybrid or balun provides the LO phase split, and the methods described in Section 6.2.3 and (6.7) through (6.9) are used to synthesize the matching circuit.

At high frequencies, LO leakage into the IF may be a concern. Since the coupling is primarily through the gate-to-drain capacitance, which is small, and the upper FETs act as switching devices, not amplifiers, the LO-to-IF isolation should be better than in a single-FET mixer. Furthermore, it is not only possible but important to connect the drains of the FETs together at the LO frequency. If the IF frequency is not too high, a capacitor can be used. The capacitive link keeps the drains of the two FETs at the same LO-frequency voltage, which, because of the circuit's symmetry, must be zero.

RF Design

Because the lower FET remains in saturation throughout the LO cycle, it effectively is an amplifier having a very low load impedance. This FET's input impedance can be found from S parameters and (6.6), and the RF matching circuit can be designed according to the methods in Section 6.2.3 and (6.7) through (6.9).

In the circuit of Figure 6.8(a), differences in the upper FETs' gate-to-source capacitances may couple some LO energy to the drain of the lower FET. If the lower FET is biased at a moderately high drain voltage, the LO energy will not cause any significant "drain-pumped" mixing in the lower FET. In Figure 6.8(b), however, there will be substantial LO voltage on the lower FET's drain. See Section 6.5.4 for an alternative.

As with the single-gate and dual-gate devices, the gate of this FET requires an IF bypass to prevent amplifier-mode gain at the IF frequency.

IF Design

The IF circuit is designed in what has become the usual manner: it should present a moderately high IF impedance to the drains of the FETs but a short circuit to LO and RF frequencies. The design of these matching and filtering structures is essentially the same as for all the previous FET mixers.

DC Bias

This requires a little thought. Remember, we said that the upper FETs were biased near pinch-off so they conduct over half the LO cycle and are turned off over the other half. Also, we stated that these devices are operated in their linear regions, as analog switches. The story is really a little more complicated.

Anyone with a little electronics experience should know that FETs in a differential configuration can't be biased that way. The reason is simple: without LO excitation, the quiescent current in the two upper FETs must equal the current in the lower FET. If we try to bias off the upper FETs, the voltage at their sources will decrease to satisfy this current condition, even if the lower FET has to drop into linear operation to do it.

When the LO is applied, however, the story changes. Now, one of the upper FETs is always on, and the lower FETs can remain in saturation, *even if the upper FETs' dc bias is at pinch-off*. If we turn on the upper FET hard enough, so we try to force more current through it than the lower FET allows, the upper FET's drain voltage will fall (again to satisfy the current condition) and it will drop into its linear region. This is advantageous, because FETs don't generate much noise when they are not current saturated, and the mixer's noise figure remains low.

6.5.4 Variations

A version of this mixer that does not require an LO balun is shown in Figure 6.8(b). In that circuit, the source node of the upper FETs is no longer a virtual ground, so to prevent imbalance, the lower FET must have a high drain-to-source resistance. This usually is not the case in GaAs MESFETs, but in MOSFETs and silicon junction FETs, the resistance is much greater and usually adequate.

6.5.5 Cautions

Avoid using the circuit in Figure 6.8 (b) with MESFETs. Although tempting, it really doesn't work well. It works much better with MOSFETs.

6.6 DOUBLY BALANCED MOSFET MIXER

This circuit is really best for low-frequency applications (say, below a couple of gigahertz). Even with MESFETs, it is not well suited for higher frequencies. When all the necessary baluns and matching circuits are included, the mixer becomes large and clumsy. At frequencies above about 20 GHz, the circuit is badly degraded by the inductance of the connections between the FETs and the imbalance caused by their inevitably unequal lengths. If you need a doubly balanced mixer at these frequencies, use diodes.

On the other hand, this mixer works very nicely in silicon integrated circuits. In an IC, an input differential amplifier can be used to reduce the mixer's unremarkable noise figure and to provide the necessary RF phase split. This results in a compact, low-power circuit.

Because it is analogous to the BJT Gilbert cell, this circuit is sometimes called a *Gilbert-cell* mixer. Purists may prefer to use this name to describe the BJT circuit alone.

Reference [8] contains a nice description of this type of mixer.

6.6.1 Characteristics

This is a true doubly balanced mixer with all the expected characteristics of such circuits: rejection of all spurious responses involving an even RF or LO harmonic, or both; LO AM noise rejection; and isolation between all three ports. In fact, the drains of the FETs are all virtual LO grounds; no LO filter or short-circuit bypass is necessary. Since it uses twice as many transistors as the differential mixer, it should have 3 dB greater output IM intercept points and 1-dB compression point. Like other active mixers, this mixer can exhibit conversion gain.

The mixer operates very much like the differential mixer in Section 6.5. It can be viewed as a combination of two such mixers, interconnected to realize a doubly balanced structure.

6.6.2 Description

The mixer is shown in Figure 6.9. The LO is applied to the upper four FETs, which, as in all doubly balanced mixers, operate as a commutating switch. The FETs normally are biased near pinch-off and remain in their linear regions when turned on. They switch the RF drain current from the lower two FETs, which remain in saturation throughout the LO cycle. The two RF FETs are, in effect, a differential amplifier.

Figure 6.9 shows resistive drain loads. *P*-channel FET current sources also can be used, but because of their low mobility, they must be large. The resulting high parasitic capacitances restrict the mixer's bandwidth. Current-source loads have the advantage of high available current at positive output-voltage peaks. They might be preferred when the maximum IF frequency is not high.

This circuit includes a current source, which removes any even-mode currents in the RF pair of FETs, even allowing them to be driven from an unbalanced source. This current source sometimes is omitted from high-frequency MESFET mixers, where it is often more trouble than it's worth.

6.6.3 Design

The design of the circuit is virtually identical to the design of the differential mixer in Section 6.5.3. The only differences are the obvious ones: the LO FETs are connected in parallel, as are the IF drains. This halves the LO input impedance,

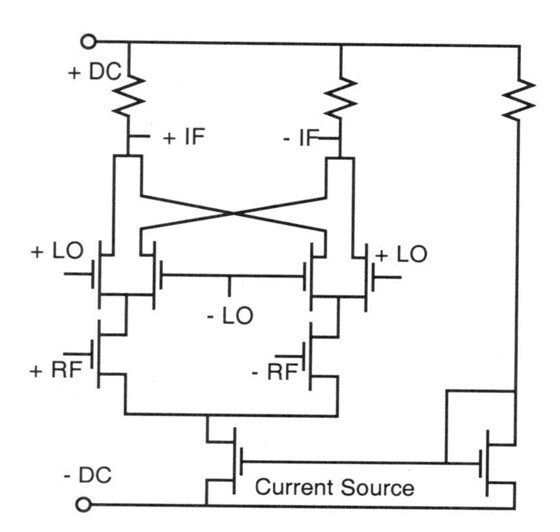

Figure 6.9 Doubly balanced MOSFET mixer. This mixer operates like a doubly balanced version of the differential mixer described in Section 6.5. The pair of FETs connected as a Wilkinson current source removes even-mode currents in the RF FETs, preserving the balance of the circuit.

compared to the singly balanced mixer, and doubles the IF load impedance at each FET. Beyond this, we refer the designer to Section 6.5.3.

6.6.4 Variations

The most important variation is the BJT Gilbert-cell mixer. This is described in Section 6.7.

A similar mixer, shown in Figure 6.10, can be realized with dual-gate devices. Although this circuit appears frequently in books and review papers, it is not used very often. Indeed, it is hard to see any significant advantage over the circuit in Figure 6.9. Like the single-device dual-gate mixer, this mixer requires IF bypassing on the LO gates; this may be difficult to implement effectively in such a complex circuit.

It is possible to interchange the LO and RF. Doing so requires a different mode of operation, with the upper FETs operated in current saturation. This configuration usually has higher noise figure than the previous one, so it generally is not the preferred one.

6.6.5 Cautions

The RF FETs tend to saturate easily on large signals. To prevent this, source degeneration (a fancy term for resistors in series with the RF FETs' sources) is

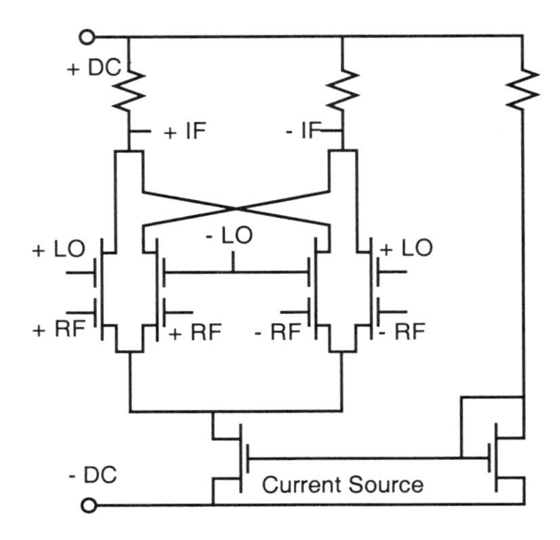

Figure 6.10 Doubly balanced MOSFET mixer using dual-gate devices.

sometimes used. Source degeneration improves large-signal characteristics slightly but increases the noise figure.

This is really a low-frequency circuit. Although it has been realized occasionally at microwave frequencies, the results have not been very good, largely because of interconnection parasitics and the difficulty of realizing three good baluns.

If you decide to operate one of these mixers at high frequencies, here are a few things to think about. We noted earlier that the drains of the LO FETs are virtual grounds for the RF and LO. This virtual ground exists only if the connections between the FETs' drains and the common node are much shorter than one wavelength; in practice, they can't be longer than 10 or 20 degrees of phase. If they become long, the connection node is still a virtual ground, but because of the length of transmission line between the node and the FET, the FET's drain is connected to a shorted stub and no longer is a virtual ground. The situation is even worse if the lengths of the lines from the node to the individual FETs' drains are unequal in length. Then, not only do we lose the virtual ground, but also the LO-to-IF and LO-to-RF isolation suffer.

Of course, the same considerations apply to the sources of the LO FETs and the lengths of the connections to the drains of the RF FETs. Also, apart from the effect on balance and isolation, the inductive reactance of these connections at high frequencies may be great enough to limit bandwidth or even cause oscillation.

Making all these lines short and equal in length is no small layout problem. Usually, it just can't be done. If the frequency is low, and the electrical lengths of the lines are short, the consequences of unequal-length interconnections are not severe; however, above 20 GHz, the problem may become very severe indeed.

6.7 GILBERT-CELL BIPOLAR MIXER

The Gilbert-cell mixer is a neat idea, but one that has become a little boring over the years. It has been just a bit too successful, and for a long time we all built them by the dozens and didn't think much about them.

Then along came the heterojunction bipolar transistor.

Now we can make these critters fly along at frequencies approaching 20 GHz. Without too much trouble, we can make multigigahertz analog multipliers and use these as direct microwave modulators, phase detectors, demodulators, and, of course, mixers. We can even do this cheaply, by using SiGe HBT technology instead of the more expensive InP and GaAs technologies. In short, they're fun again.

6.7.1 Characteristics

The basic circuit is shown in Figure 6.11. Like the very similar MOSFET mixer, the Gilbert cell is doubly balanced. It has all the delightful properties of a doubly balanced mixer, and, in a sense, more. Bipolar-transistor technologies are mature, and it is not difficult to make very well matched BJTs. This results in mixers having

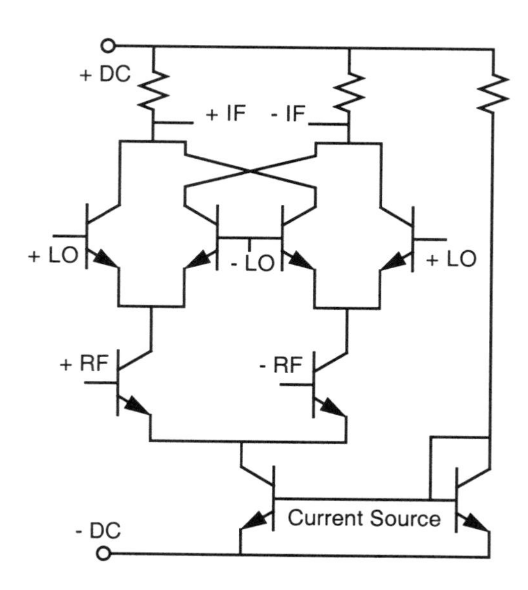

Figure 6.11 Bipolar Gilbert-cell mixer.

an extraordinary degree of balance; for example, at frequencies below 1 GHz, LO-to-IF isolations can be upward of 50 or 60 dB. In view of the inherently low port isolations of active mixers, this is especially surprising.

Gilbert cells have a number of worthwhile uses. As analog multipliers, they usually are "linearized" so the output current is equal to the product of two input currents. For most mixer applications, this is not necessary, although it may be helpful for modulators and similar signal-processing circuits.

6.7.2 Description

Figure 6.11 shows the circuit. In a mixer, the upper four devices, to which the LO is connected, are driven hard enough to turn them alternately off and on over the LO cycle. These devices then operate as commutating switches, while the lower (RF) devices remain in linear operation[4] over the whole LO cycle. The RF BJTs operate very much like a differential amplifier. As such, the emitters of the RF devices are a virtual ground for the RF signal, and the emitters of the LO devices are a virtual ground for the LO. There is ideally (and in practice) no LO voltage on the collectors of the RF devices.

Figure 6.11 shows collector load resistors. Active loads, consisting of *pnp* BJTs, also could be used; such devices may have to be large, and their parasitic capacitances are likewise large. The trade-off between active and passive loads depends on the characteristics of individual fabrication technologies.

Many Gilbert-cell mixers and multipliers use emitter degeneration (series resistors) to limit the gain of the differential stage (the devices driven by the RF signal in Figure 6.11). This improves the large-signal handling, and probably the linearity, at some expense to the noise figure.

A classical Wilkinson current source provides the current to the differential pair.

6.7.3 Design

The design philosophy is similar to that of the differential mixer in Section 6.5.3 and the MOSFET mixer in Section 6.6.3. Again we note that the emitters of the upper devices are virtual grounds for the LO and the emitters of the lower devices are virtual grounds for the RF. This allows us to view the input impedance at the base as simply the same input impedance as a single, common-emitter device would have. From this observation, we obtain the input impedances for designing matching networks.

The bases of the RF devices should be shorted at the IF frequency. This prevents amplification of IF noise, which could increase the noise figure significantly. There

---

4. FETs are in their *linear* region of operation at low drain voltage and are *saturated*, or active, at higher voltages. BJTs are said to be *saturated* at low collector voltage and in *linear*, or active operation at higher voltages. This terminology is more than a little confusing, we admit, but, still, it is accepted terminology, and we're stuck with it.

is no need to provide LO or RF short circuits to the drains of the LO devices; they are virtual grounds.

6.7.4 Cautions

One of the greatest frustrations in designing Gilbert-cell mixers is that the descriptions of these circuits in most general electronics textbooks focus on mid-frequency performance. By *mid-frequency*, we mean frequencies where the transistor's parasitic capacitances are negligible; we RF and microwave folks might call this *resistive operation*. Unfortunately, one of the fundamental dividing lines between us high-frequency people and the folks who deal in "general electronics" is that they operate at mid frequency, while we operate at high frequencies. More often than not, we operate FETs and BJTs at frequencies where they are just barely capable of doing something useful, be it amplifying, oscillating, or mixing. Unfortunately, most of the conventional wisdom about these circuits comes from the mid-frequency assumption. Much of this is invalid at RF and microwave frequencies.

For example, one of the ostensible benefits of the HBT is its high dc collector-to-emitter resistance. This should make it ideal as a current-source transistor. However, at high frequencies, the collector-to-emitter impedance of an HBT often is surprisingly low. This is caused by feedback through the collector-to-emitter capacitance combined with the HBT's inherently high transconductance. It is surprising that a feedback capacitance of a few femtofarads could make such a dramatic change in output impedance, but it does. Keeping the base well-grounded helps minimize this effect but does not eliminate it entirely. The moral of the story is simple: be careful of your assumptions.

REFERENCES

[1] Maas, S. A., *Nonlinear Microwave Circuits*, New York: IEEE Press, 1997.

[2] Kollberg, E., ed., *Microwave and Millimeter-Wave Mixers*, New York: IEEE Press, 1984.

[3] Kurita, O., and K. Morita, "Microwave MESFET Mixer," *IEEE Trans. Microwave Theory Tech.*, Vol. MTT-24, June 1976, p. 361.

[4] Matthaei, G., L. Young, and E. Jones, *Microwave Filters, Impedance-Matching Networks, and Coupling Structures*, Norwood MA: Artech House, 1980.

[5] Tsironis, C., R. Meirer, and R. Stahlmann, "Dual-Gate MESFET Mixers," *IEEE Trans. Microwave Theory Tech.*, Vol. MTT-32, March 1984, p. 248.

[6] Sevick, J., *Transmission-Line Transformers*, Atlanta: Noble Publishing, 1996.

[7] Maas, S. A., *Microwave Mixers*, 2nd ed., Norwood, MA: Artech House, 1993.

[8] Sullivan, P., W. Ku, and B. Xavier, "Active Doubly Balanced Mixers for CMOS RFICs," *Microwave J.*, October 1997, p. 22.

Chapter 7
FET Resistive Mixers

Many of us in our childhood delighted in turning ordinary objects into unique and special creations. Through our imaginations, a spoon became an airplane, a bent stick became a gun, and the neighborhood pussycat became a ferocious lion. Most of us outgrew this, but a few of us didn't. As proof, in this chapter an ordinary, unbiased MESFET will become a mixer.

FET resistive mixers are designed for low-distortion applications. They exhibit very low levels of both IMD and spurious responses, and their 1-dB compression points are very high. Other performance characteristics—noise, conversion loss, and LO power—are similar to those of diode mixers. Low IMD is not their sole purpose, however; these mixers have a number of other delightful properties, including low sensitivity to AM noise on the LO and low $1/f$ noise. They also are a convenient solution to the problem of realizing mixers in certain monolithic technologies where Schottky-barrier diodes do not exist.

Like diode mixers, FET resistive mixers can be designed as single-device circuits, singly balanced circuits, and doubly balanced circuits.

Table 7.1 lists the mixers described in this chapter. They include both single-device and balanced structures and include circuits appropriate for both RF and microwave applications.

7.1 FUNDAMENTALS OF FET RESISTIVE MIXERS

FET resistive mixers [1–3] are an approximation of an ideal known as *linear mixing*, mixing by a linear resistance. Such mixing, if it actually could be achieved, is theoretically capable of distortionless frequency conversion. Although perfection, especially in electronics, is a scarce commodity, the FET resistive mixer comes as close to true linear mixing as anything invented so far.

We begin with a discussion of resistive mixing, show an example of an ideal linear mixer, and finally discuss linear mixing in unbiased (or *resistive*) FETs. Finally, we show how such mixers are realized in practice.

Table 7.1 Mixers Described in This Chapter

Mixer Type	Characteristics	Typical Applications
Single-device FET resistive mixer	A good circuit for general use. Requires an LO-IF diplexer.	General applications; lack of an IF balun or hybrid makes this mixer somewhat more practical for ICs. Rejects AM LO noise. A nice circuit for RF ICs.
Singly balanced 180-degree mixer	Lower distortion, automatic drain short-circuit for the LO. Requires an IF balun or 180-degree hybrid. Requires filtering to separate the RF and IF.	Especially useful when the IF is low enough to allow a wire-wound transformer for the output balun. In low-frequency mixers using transformers, the IF and RF may overlap.
Doubly balanced ring mixer	Very low distortion but difficult to design optimally at high frequencies because of large layout parasitics. Bandwidth is limited by hybrids.	A very good circuit for general wireless, RF, and lower microwave applications.

7.1.1 Linear Mixing

In Chapter 3, Equations (3.1) - (3.6), we saw that pumping a Schottky-barrier diode with a large signal turns its junction into a time-varying, small-signal conductance. This time-varying conductance then mixes with the applied RF voltage, resulting in mixing products involving the RF frequency and the conductance waveform's fundamental frequency and harmonics. The key equation in this derivation is

$$i(t) = g(t)v_s(t) \tag{7.1}$$

where $g(t)$ is the conductance waveform, $v_s(t)$ is the applied voltage, and $i(t)$ is the resulting current.

It is critical to note that (7.1) is a *linear* equation. The process of frequency conversion, which it describes, is a linear one, and an ideal mixer is a linear component. True, we normally must use nonlinear devices to realize mixers, and this results in distortion, harmonic generation, spurious responses, and other nonlinear phenomena. Nevertheless, (7.1) show us that there is no fundamental reason why a nonlinear device should be necessary to produce frequency mixing. We do, however, need the time-varying conductance. Do linear, time-varying conductances exist? If so, we can use one to create a distortionless mixer.

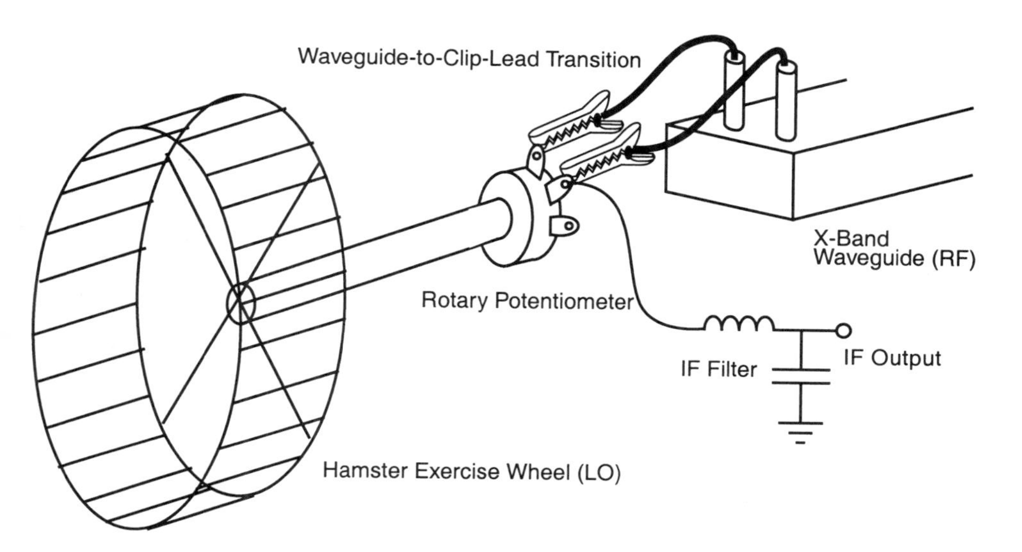

Figure 7.1 A linear mixer made from a hamster exercise wheel and a rotary potentiometer. The stops have been removed from the potentiometer so it can rotate freely. The exercise wheel (once the hamster has been introduced and properly motivated) is the LO, and the potentiometer provides the time-varying conductance. The RF excitation comes from the X-band waveguide.

7.1.2 Our First Linear Mixer: The Hamster-Pumped Mixer

Figure 7.1 shows one way to make a linear X-band[1] mixer using a hamster's exercise wheel and a rotary potentiometer. The stops must be removed from the potentiometer so it can rotate freely through 360 degrees. When the hamster is introduced, (and, mind you, this must be a very fast hamster), the running hamster turns the wheel, the wheel turns the potentiometer, and the potentiometer becomes a time-varying conductance. If we can make the hamster turn the wheel fast enough, at least $8 \cdot 10^9$ RPM, we can achieved X-band mixing. We only need to couple the RF signal to the potentiometer and extract the IF. (We recognize that the approach used in Figure 7.1, especially the X-band waveguide-to-clip-lead transition, may need a little more thought. We're just trying to illustrate the principle.)

We don't really recommend making a mixer this way—for one thing, it's awfully difficult to phase-lock the LO—but the circuit, however impractical, shows that frequency mixing does not require a nonlinear device. Furthermore, this is a *linear* mixer: it has no IMD or spurious responses involving a harmonic of the RF frequency. If the potentiometer's resistance profile were to vary as the sine of the

1. X band is the frequency range from 8 to 12 GHz.

angle of shaft rotation, it wouldn't even have any spurious responses involving LO harmonics.

7.1.3 An Improved Linear Mixer: The FET Resistive Mixer

How do we make a more practical mixer than the hamster-pumped mixer of Section 7.1.2? First, we need to replace the hamster with a better device. This will improve both the mixer's reliability (a hamster's mean time to failure is only about 2 years) and the LO's frequency stability. Our new device must have a linear resistance that can be controlled by an LO voltage or current.

The channel of an unbiased FET meets these requirements. At low drain voltages, the channel operates as a linear resistance. It is simply a slab of semiconductor, whose small-signal conductance, G, is

$$G = \frac{\mu q N_d W t}{l} \tag{7.2}$$

where μ is the electron mobility, q is electron charge, N_d is the channel's doping density (which, for now, we assume to be uniform), W is the channel's width, t is the thickness of the undepleted portion of the channel, and l is its length. Varying the gate voltage varies the gate depletion depth, thus the channel's thickness t, and therefore the conductance. In most microwave or RF FETs, the conductance can be varied from zero (at or below pinch-off) to a few tenths of one siemen. Because this resistive characteristic is used for mixing, in contrast to the transconductance mixer of Chapter 6, we call this a *FET resistive mixer.*

Mixer Circuit

Figure 7.2 shows the circuit of a single-device FET resistive mixer. The LO is applied to the gate via a matching network, which is not significantly different from the type of gate-matching circuit used in an active mixer. Dc gate bias is important for optimizing the mixer's performance; it is applied to the gate through a conventional bias circuit. At the drain, a simple diplexer separates the RF and IF signals. An RFC short-circuits the drain at dc; if a dc path to ground exists in the diplexer, as is often the case, an RFC is not necessary.

The filters in the RF diplexer are critical to the performance of the mixer. They must short-circuit the drain at the LO to prevent pumping of the drain by LO energy coupled through the gate-to-drain capacitance. This concern is real: although the gate-to-drain capacitance of a biased FET may be very low, it is not low when the drain bias is removed. At zero drain bias the gate-to-channel capacitance is roughly evenly divided between the gate-to-source and gate-to-drain capacitances, and the gate-to-drain capacitance is several times its value in a drain-biased device.

The LO circuit is not especially critical. Although it is prudent to short circuit the

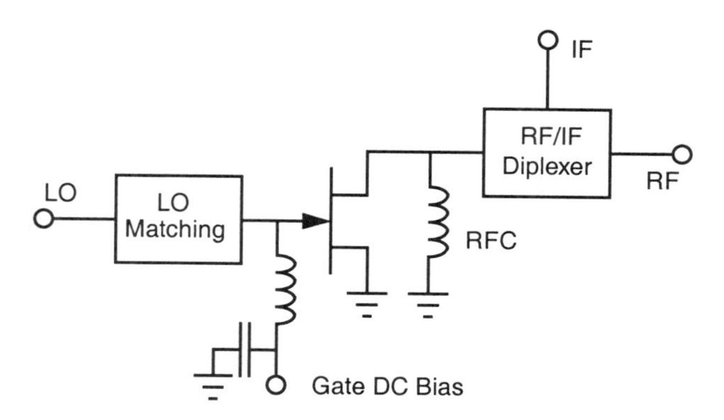

Figure 7.2 A single-device FET resistive mixer. The RFC keeps the drain-to-source voltage at zero. It may not be necessary if a drain-source dc short circuit exists in the diplexer.

gate at the RF, very little of the small-signal RF voltage is coupled to the gate, and little pumping of the gate by the RF can be expected.

The RF input and IF output impedances of the FET are surprisingly accommodating. When an ordinary small-signal MESFET is used, the RF input impedance is approximately 100Ω in parallel with the gate-to-drain capacitance. (Here we are assuming that the gate is shorted at the RF frequency.) As with a diode mixer, the IF output impedance is somewhat higher, perhaps 150–200Ω. Usually, the gate-to-drain capacitance is negligible at the IF frequency in a downconverting MESFET mixer.

The LO input impedance, which is determined by the gate-to-source capacitance and parasitic resistances, is pretty much the same as in an active mixer or frequency multiplier. The method for determining this input impedance described in Section 6.2.3 is not applicable here, because the S parameters of an unbiased device usually are unavailable. (If they are available, the input reflection coefficient is simply S_{11}.) A better estimate for the LO input impedance at the gate, Z_{LO}, is

$$Z_{LO} \cong R_s + R_g + \frac{1}{j\omega_{LO} C_{gc}} \tag{7.3}$$

where R_s is the FET's source resistance, R_g is its gate resistance, C_{gc} is the gate-to-channel capacitance, and ω_{LO} is the LO frequency. We have not included the FET's intrinsic resistance, R_i, because that element does not apply to the unbiased FET.

Channel Conductance Waveform

Like the conductance waveform of a diode or the transconductance waveform of a

FET used in an active mixer, the conductance waveform of a resistive FET should be adjusted to maximize its fundamental-frequency component. To do so, we bias the FET near pinch-off and pump it as hard as possible. This results in low conversion loss and practical input and output impedances.[2] Operating the FET mixer in this manner will indeed result in very good performance, in terms of conversion loss. However, it may not result in the lowest possible IMD.

To achieve low distortion, we must recognize a couple of facts about distortion in mixers. First, most distortion is generated during the brief periods when the LO voltage is in the most strongly nonlinear portion of the FET's I/V characteristic. This occurs while the gate voltage is near pinch-off and the FET is turning on or turning off. For lowest distortion we want the FET to turn on and off as fast as possible, thereby spending as little time as possible at gate voltages where the channel is most strongly nonlinear. To do this, we need the largest possible LO voltage at the gate, so, in general, high LO power provides low distortion. LO voltage cannot be increased indefinitely, however, because the gate begins to rectify on positive peaks. The resulting nonlinearity increases the mixer's distortion; the cure is to reduce the gate dc bias (that is, make it more negative) to allow stronger LO pumping. This narrows the FET's conductance pulse, resulting in higher RF input and IF output impedance. Eventually, as the gate bias is reduced, the FET's gate starts to break down on negative LO-voltage peaks, and the LO voltage can be increased no further.

The higher impedance can be ameliorated somewhat by increasing the gate width of the device. Not only does this reduce the impedances, but a larger device simply can handle greater power; for these reasons, power devices often are used for FET resistive mixers. Wider devices require higher LO power, however, and have greater capacitive parasitics. In general, it is best to use the widest device possible, within the limits of available LO power and matching.

Another way to speed the transition between the FET's on and off states is to short circuit the gate at LO harmonics. This maximizes the rate at which the gate-to-channel capacitance can charge and discharge. Since this capacitance is only weakly nonlinear, the effect is not great, but may be of some help in minimizing distortion.

FET Model

The conventional equivalent circuit for a FET used in a resistive mixer is not the same as the equivalent circuit for an active FET. (The active MESFET's equivalent circuit, for example, is shown in Figure 2.18.) A more appropriate equivalent circuit

---

2. There is a large body of literature, dating back to the 1940s, that describes the effect of a resistive element's conductance waveform on conversion loss. The channel-conductance waveform that we obtain by biasing the FET in this way certainly is not optimum in this sense; that is, it does not provide the absolute minimum conversion loss. It is "optimum" in a practical sense, however, because it provides conversion loss that is as low as can be achieved in practice.

is shown in Figure 7.3. It represents a MESFET, but the equivalent circuit of a JFET or MOSFET is similar.

In deriving the circuit of Figure 7.3 from Figure 2.18, a number of parasitic elements have been eliminated. The elements R_{ds} and C_i in Figure 2.18 model the effects of traps in the FET's channel when it is biased into current saturation and the channel electric field is strong. This is not the case here, so these elements are not included. C_{ds} is a small capacitance, the combination of intermetallic parasitics and the slightly capacitive behavior of the biased channel's dipole layer. This capacitance is much smaller when the device is unbiased; in any case it invariably is negligible compared to the high average conductance of the resistive channel. Finally, R_i, in the biased device, models the extra resistance between the gate and the source in biased FETs. This results largely from the asymmetrical shape of the depletion region, caused by the large voltage gradient along the channel. Again, in resistive devices this situation does not exist, so R_i has been omitted from Figure 7.3.

How do we model the current source in a resistive FET? Most of the existing FET *I*/*V* models have been designed for active devices, not resistive devices. Nevertheless, most of these models are useful for resistive FETs as long as the model's parameters are adjusted to reproduce the measured *I*/*V* characteristics in the vicinity of zero drain voltage, and the model does not misbehave when the drain voltage is negative. It is a common error to use a model that has been derived for an active device and may not be accurate at a dc drain voltage of zero.

The JFET and MOSFET equivalent circuits in Chapter 2 are much simpler than the MESFET, and the same simplifications can be applied to Figure 7.3. For example, the gate resistance often is negligible in low-frequency JFETs and MOSFETs, so that parameter can be eliminated. Similarly, the diodes that model the Schottky junction in Figure 7.3 should be removed for MOSFETs.

Finally, we must emphasize the fact that this model is not adequate for predicting IMD levels of FET resistive mixers. Modeling a FET for this purpose is an especially

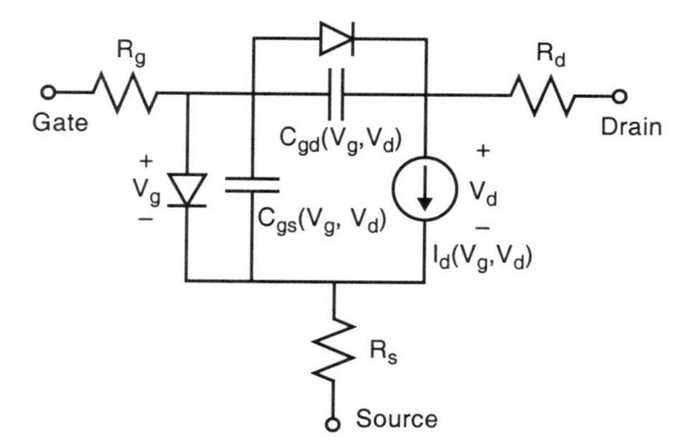

Figure 7.3 The large-signal equivalent circuit of a FET used in a FET resistive mixer. This circuit is derived from the one in Figure 2.18.

difficult problem because the weak nonlinearity of the FET's channel is much more difficult to model accurately than a strong nonlinearity. Still, the lack of an appropriate IMD model is not too limiting. Empirical evidence shows that a mixer designed according to the criteria in the following section will achieve optimum IMD performance, as well as optimum noise figure and conversion efficiency.

Design Requirements for FET Resistive Mixers

Let's summarize what we need to do to realize a good FET resistive mixer. These should sound pretty familiar by now:

1. Short-circuit the drain to the source at the LO frequency. Because the device is not active, a drain short circuit at harmonics is not necessary.
2. Short-circuit the drain to the source at dc. It is essential to keep the dc drain voltage at zero. If no such short exists, the drain can become biased by rectification in the gate-to-channel junction.
3. Match the gate, if possible, at the LO frequency. Because of the FET's high input Q, it may not be possible to achieve a good match over even a modest bandwidth. The technique for doing this is the same as for an active mixer; see Section 6.2.3.
4. Match the drain at the RF and IF frequencies.
5. Adjust the FET's gate bias halfway between breakdown and conduction. This will be somewhat below (i.e., more negative than) the pinch-off voltage.
6. Adjust the LO level just short of generating dc gate current.

As with balanced diode and active FET mixers, some of these requirements, especially the drain short circuit for the LO, can be met automatically. Of the rest, the LO, RF, and IF matches are what most designers would do in any case, and short circuiting unwanted frequency components at all ports has become the general rule for all nonlinear circuit design.

Performance

The performance of a well-designed FET resistive mixer depends on frequency, the type of device, and the type of mixer (single-device or balanced). In general, the conversion loss should be about the same as that of a diode mixer at the same frequency or, because of the FET's greater parasitic capacitances, slightly higher. The output intermodulation intercept points of all products should be much higher than those of a diode mixer; at the same LO level, an improvement of 10–15 dBm in third-order intercept point is not unusual. Most dramatic is the improvement in the mixer's one-dB gain compression point; this typically is 10 dB better than a diode mixer, although the IMD level may increase more rapidly than expected as gain compression is approached.

Because of substantial shot noise in its diodes, a diode mixer's single-sideband noise figure usually is at least a few tenths of a dB greater than its conversion loss. In a FET resistive mixer, the noise is almost entirely thermal, so the noise figure usually is precisely equal to the conversion loss.

Because the gate of the FET is reactive, FET resistive mixers do not require much LO power. Although the LO power absorbed by the FET may be very low, the well-known limitations in matching reactive devices over a bandwidth largely determine the LO power requirements. Over moderate bandwidths, the required LO power, for minimum IMD, rarely is more than 10 dBm per FET; often it is much lower. When the gate is conjugate matched, the required LO power, P_{LO}, is

$$P_{LO} = \frac{1}{8}(V_{g,\,max} - V_{g,\,min})^2 \omega_p^2 C_{gs}^2 (R_g + R_s) \tag{7.4}$$

where $V_{g,max}$ is the maximum gate voltage, established by the point where the gate starts to conduct appreciably; $V_{g,min}$ is the minimum gate voltage, limited by breakdown; ω_p is the LO frequency; and the other terms are defined in Figure 7.3. Remember, (7.4) is a lower limit; it is valid for a conjugate-matched gate only, which is possible only in special cases. In most cases there will be at least 2–3 dB of reflection loss.

This type of mixer has very low downconversion of LO AM noise. In diode mixers or active FET mixers, the LO is applied to the mixing element along with the RF signal, so the LO AM noise is no different from an RF excitation. The noise is simply downconverted along with the RF, increasing the mixer's output noise, and balanced structures are needed to reduce LO noise. In the FET resistive mixer, however, the LO and its AM noise are applied to the gate, and the conversion efficiency between the gate and IF is very low. For this reason, even single-ended FET resistive mixers usually do a good job of rejecting AM LO noise. (Alas, this characteristic does not apply to LO *phase* noise; LO phase noise modulates the downconverted waveform degree for degree, as in any other mixer.)

The $1/f$ noise in these mixers is low as well. The lack of high currents through the mixing device gives the FET resistive mixer very low levels of $1/f$ noise, making it ideal for use in mixers that downconvert directly to baseband. See [4] for an example.

7.2 SINGLE-DEVICE MIXER: RF APPLICATIONS

7.2.1 Characteristics

This mixer, shown in Figure 7.4, is appropriate for use at frequencies up to a few gigahertz. At higher frequencies, coupling through the gate-to-channel capacitance, combined with the limited Q of lumped elements (whether realized by chip components or IC structures) is not adequate to reduce LO leakage to a tolerable level. Because of its good performance, low LO-power requirements, and

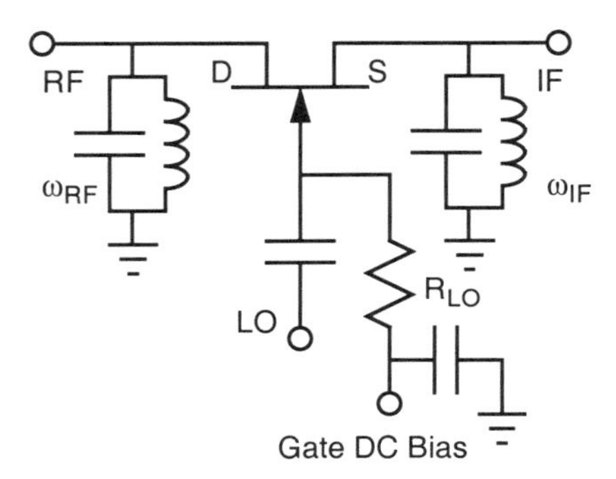

Figure 7.4 A single-FET resistive mixer appropriate for use at frequencies up to a few gigahertz. It is appropriate for use in either monolithic or discrete hybrid technologies.

practicality for use in monolithic circuits, this circuit does well in small RF and wireless receivers. Additionally, this circuit is instructive: it neatly achieves all the requirements listed in Section 7.1.3.

7.2.2 Description

The mixer essentially is a series analog switch tuned at the RF input and IF output. In order to keep the circuit simple, the gate is driven directly from its 50Ω LO source. To improve the LO VSWR, at some cost to the required LO level, a 50Ω resistor is connected to the gate. This resistor is also used as a path for the gate's dc bias.

The RF and LO ports have simple, L-C parallel tuned circuits. These ground the drain and source terminals at dc and, if their Qs are high enough, at the LO frequency as well. A well-designed circuit has drain-to-source voltage components only at the RF and IF frequencies.

The resonators decouple the RF and IF nicely, so the FET's channel is terminated in 50Ω at both the RF and IF frequencies. In a monolithic circuit the size of the device can be adjusted to achieve good RF and IF VSWRs with these terminations. In discrete circuits the inductors of the RF and IF resonators can be tapped to adjust the input impedance.

At frequencies where this type of mixer is useful, the FET's gate-to-channel capacitance usually is negligible. For example, the ~0.3 pF gate-to-channel capacitance of an ordinary MESFET has a reactance of over 200Ω at 2.5 GHz, a negligible value compared to the net 25Ω resistance (50Ω source and 50Ω R_{LO}) in shunt with the gate.

The 50Ω gate resistor increases the LO power requirement to a value 6 dB greater than a circuit having no resistor. (Why 6 dB? The gate requires a certain ac voltage to pump the device. The resistor R_{LO} halves the applied voltage, so the LO source voltage must double to compensate. This doubling of voltage requires a 6-dB increase in available LO power.) Furthermore, the 50Ω source probably is not optimum for minimum LO power, so this 6-dB penalty is placed on top of the penalty for nonoptimum design. Still, even with these penalties, the LO power requirement is likely to be little more than 10 dBm, not a terribly frightening value for most applications. In cases where this level is still too great, see Section 7.2.4 for alternatives.

The FET's gate-to-channel capacitance limits the frequency at which this circuit is useful. LO leakage through this capacitance into the RF port usually is the determining factor. Often the RF and LO are not very far apart in frequency, and the Q of the RF tuned circuit cannot be made high enough to reject LO leakage significantly. Even when the frequency separation is great, the limited Q of the inductor and capacitor may prevent adequate isolation.

The RF and IF ports do not require blocking capacitors, but the LO port does require one. A bypass capacitor is needed in the dc bias circuit.

7.2.3 Design

Device Selection

As we indicated in Section 7.1.3, the largest device (greatest gate width) possible should be used. As the device increases in size, port impedances decrease, although not necessarily in proportion to device size, and LO-to-RF and LO-to-IF leakage increases. As long as the gate capacitance is negligible at the LO frequency, LO power requirements do not change; however, if a more conventional matching circuit is used at the gate, LO power requirements may increase.

To minimize LO power, the FET's pinch-off voltage must be as close to zero as possible; usually a pinch-off voltage of about 1V, typical of small-signal FETs, is fine. The low pinch-off voltage allows the FET to be turned on and off with minimal ac gate voltage, thus minimizing LO power. Unfortunately, most large devices are designed for power-amplifier circuits, where a high pinch-off voltage is preferable. Accordingly, large devices usually have high pinch-off voltages.

Trading off device characteristics, LO power, and leakage usually requires nonlinear analysis.

Enhancement-mode MOSFETs work nicely in this circuit at frequencies below about 2 GHz. One important advantage is that their pinch-off voltage (which in MOSFETs is called the *threshold voltage*) is somewhat above 0V. The optimum gate bias in a FET resistive mixer is somewhat below pinch-off, so in enhancement-mode MOSFETs, zero-voltage gate bias is nearly optimum. MOSFETs have two other advantages leading to low distortion: their channels are silicon, which is more linear than GaAs, and, because of the lack of gate rectification, they can be pumped very hard. Unfortunately, MOSFETs also have negative characteristics in the quest for high performance: their generally higher minimum channel resistance, which limits conversion efficiency; slower switching; and greater parasitic capacitances. It is not clear, at this point, whether GaAs or silicon MOS devices have superior IMD performance at lower frequencies. Because of their lower minimum channel resistances, GaAs devices clearly have lower conversion loss at all frequencies and are superior in all respects above approximately 2 GHz.

HEMTs probably are a last choice for FET resistive mixers. Although their capacitive parasitics are small and only very low LO voltages are needed to turn them on, the channel resistance of a HEMT is much more strongly nonlinear than other types of FETs. HEMTs are best used when very little LO power is available or for millimeter-wave mixers, where the importance of low distortion is secondary.

Design

It is a good idea initially to design an ideal circuit and to simulate it by nonlinear analysis. This circuit is used to determine the best device size and port impedances. Finally, a detailed design finishes the process. Here is the approach:

1. Create a circuit like Figure 7.4, using ideal elements, but leave out R_{LO}. Use a large-value inductor (RFC) for the bias circuit. Make the capacitance in the L-C resonators very large. This minimizes their bandwidths and guarantees that the drain and the source are grounded at the LO frequency.
2. Analyze the circuit and determine the port impedances and conversion loss at a single RF and LO frequency.
3. Adjust the device size by scaling (in the case of an IC) or by selecting different devices (in the case of a discrete circuit) until both the RF and IF input and output impedances are within a factor of two of the desired value. For each device size, adjust the LO power and dc bias to optimize the performance. You will find that you can get good conversion loss and port impedances with almost any device at some bias voltage, but unless the bias is below pinch-off, optimum IMD performance will not be achieved. Therefore, keep the gate bias at or below pinch-off.
4. Note the LO input impedance and port impedances when the performance looks good.
5. Now design the mixer using real components instead of ideal ones. Include all the parasitics for the inductors and capacitors used in the resonators. Include R_{LO}, if low LO VSWR is more important than low LO power. The capacitance in each resonator is found from the loaded Q:

$$Q = \frac{\omega_0 C + Im\{Y\}}{Re\{Y\} + G} = \frac{\omega_0}{\Delta\omega} \tag{7.5}$$

where ω_0 is the center frequency of the RF or IF band; Y is the RF or IF input or output admittance of the FET, as calculated by nonlinear analysis; G is the port admittance ($G = 1/R$, where R usually is 50Ω); and $\Delta\omega$ is the bandwidth. Solve (7.5) for C. The inductance is

$$L = \frac{1}{\omega_0(\omega_0 C + Im\{Y\})} \tag{7.6}$$

that is, we have absorbed the reactive part of the input or output admittance into the resonator.

6. If the real part of the input or output impedance determined in step 3 is too high, tap the inductor to help match the input or output port. Remember that this will change the loaded Q, as well as the conversion loss, and you may have to modify L and C to compensate.
7. Finally, use nonlinear analysis to calculate the performance of the complete mixer. Especially note the LO-to-RF isolation and make certain it is adequate. Similarly, confirm that the RF and LO circuits are on frequency and have adequate bandwidth. The easiest way is to view the input and output admittances; the band center is the point where the real part is zero.

Tuning and Adjustment

The tuning adjustments are the obvious ones: LO power, dc bias, and the two L-C resonators. The conversion efficiency should be relatively sensitive to gate bias at low LO levels, and even at very low LO levels it should be possible to achieve good conversion efficiency. At low LO power the IMD will be high. As LO level is increased, the optimum dc bias point also should become more negative, and conversion loss should become less sensitive to bias. In MESFETs, at the optimum dc bias voltage and LO level, there should be no gate current, but adjusting the gate bias a few tenths of a volt either positive or negative should cause gate current to appear.

7.2.4 Variations

Reduced LO Power

The increase in LO power caused by the presence of R_{LO} is unacceptable in many applications. If so, the best way to design the LO is to use the method described in Section 6.2.3 and (6.7) to (6.9). That method consists of resonating the LO input capacitance with a series inductor and adjusting the source impedance to achieve a Q that provides just enough LO bandwidth. A transformer or other conventional means can be used to adjust the source impedance. If this approach still does not provide an acceptable LO design, which often is the case when the bandwidth is great, a conventional L-C matching circuit can be used. See Reference [14] of Chapter 1 for more information, and be prepared to spend some time optimizing the circuit.

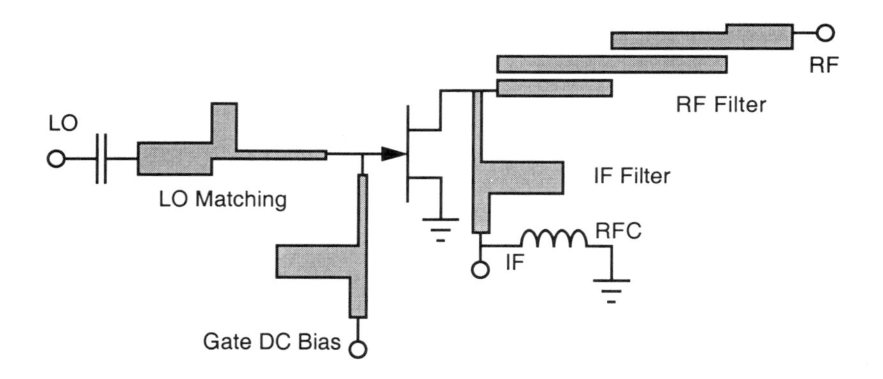

Figure 7.5 A microwave implementation of the FET resistive mixer uses distributed circuits for the RF, IF, and LO matching circuits and the gate bias circuit. The capacitor at the LO port is a dc block, and the RFC keeps the drain dc voltage at zero.

Microwave Single-Device Mixer

The circuit in Figure 7.4 is a little clumsy to realize as a microwave circuit. A better microwave version is shown in Figure 7.5. There are a number of ways to realize such circuits; this is only one example.

The circuit uses a coupled-line RF filter. This filter has good performance and is ideal for bandwidths around 20% of the RF center frequency [5]. Most important, its output impedance is high at frequencies well below the passband, so it does not load the IF. The IF filter is a simple structure consisting of a series high-impedance line and a shunt low-impedance stub. The lengths of these elements are one-quarter wavelength at the RF center frequency. Their impedances are chosen so the structure behaves as a series-L, shunt-C filter at the IF. The RFC provides a dc ground to the drain; it may not be necessary if this function is provided elsewhere in the IF.

The gate-matching circuit is a series-L, shunt-C structure. This usually is adequate to match the gate adequately over a 10% to 20% bandwidth; remember, the input Q is much lower at microwave frequencies than at RF. Some degree of empirical tuning often is necessary to compensate for the discontinuities in the structure and uncertainties in the FET's input impedance. The gate-bias circuit is similar to the IF circuit.

The design of this mixer is straightforward. The RF and LO input impedances and the IF output impedance at the device can be found in essentially the same way as those of the low-frequency circuit. In this circuit, however, there is limited provision for matching the RF and IF ports. It is best to select the device and gate bias to achieve an adequate match. This is something of a juggling act; the bias and the LO level must be chosen simultaneously to match the device and still provide the optimum conditions described in *Tuning and Adjustment*, on page 225. Another possibility is to use a more complex set of filters that include some degree of matching as well as filtering.

7.2.5 Cautions

This is a relatively easy circuit to get working, and few problems should be expected. One possible problem is a strange oscillation that occurs only when LO excitation is applied. It is especially evident in two-tone tests and behaves like IMD components whose level is very sensitive to dc gate bias. This type of oscillation is caused by a combination of the nonlinear gate-to-channel capacitance and some inductance in the circuit or perhaps by interaction with the dc-bias power supply. The solution in the former case is to load the gate with extra resistance, especially if R_{LO} is not used. In the latter case, add series resistance and a large capacitor to the bias circuit.

7.3 180-DEGREE SINGLY BALANCED FET RESISTIVE MIXER

7.3.1 Characteristics

This is a true singly balanced mixer. Its conversion loss is about the same as the single-device mixer, but its odd-order IPs are 3 dB greater. Because of its inherent rejection of even-order IMD, its even-order IPs are typically 20 dB better, often more. This makes it especially nice for applications where even-order IMD is a problem.

Because there is no RF balun and a matching circuit is usually unnecessary, the RF bandwidth can be very broad. Since the RF port drives two FETs in parallel, the input impedance usually is very close to 50Ω.

7.3.2 Description

One of two possible configurations for a singly balanced mixer is shown in Figure 7.6. In this circuit, the LO is applied 180 degrees out of phase at the gates and the RF is applied in phase. The alternative is to apply the RF out of phase and the LO in phase. We prefer the former circuit, because it is a bit more practical: it has lower RF loss, thus better noise figure, and the transformer can be used to optimize the LO source impedance, thus minimizing LO power. The FETs are in parallel at the RF, and this usually results in a good RF match. Most important, in this configuration the FETs' drains are a virtual ground for the LO, so the optimum LO termination is provided automatically and the LO-to-RF isolation is very good.

One tempting circuit to avoid is the configuration where both the RF and the LO are applied out of phase. This configuration seems attractive because the IF currents in the two devices are in phase, so no IF balun is needed. The advantage is obvious: eliminating the IF balun saves space. The disadvantage, however, is serious: the

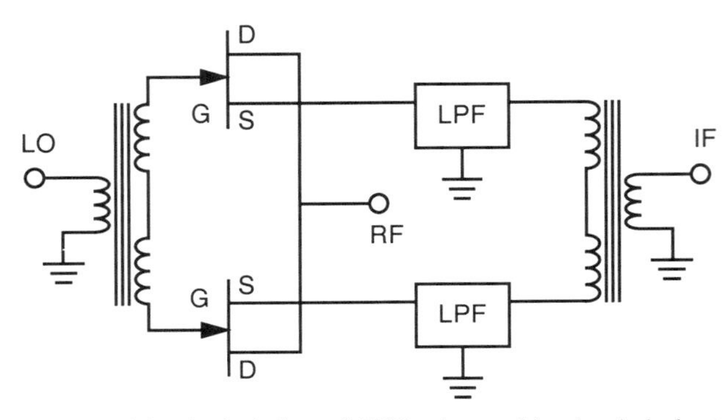

Figure 7.6 Of several possible singly balanced FET mixers, this circuit is best for virtually all applications. Its main advantage is the IF and LO virtual ground at the FETs' drains.

circuit is simply two single-FET mixers in parallel, not a true balanced mixer. A way to avoid the IF balun in a true singly balanced mixer is shown in Section 7.3.4, but for reasons given in that section, this is not a very good circuit either. Keep the IF balun.

In the circuit in Figure 7.6, the RF current returns through the low-pass filters. It is tempting to make it return through a center tap in the transformer and to eliminate the filters, but this isn't a good idea: it would require the IF transformer bandwidth to include the RF band, and this inevitably would compromise the IF performance. It also would require avoiding the better transformers—transmission-line types—which do not have a center tap. A less obvious problem is that LO leakage likely would be within the transformer's bandwidth, and the transformer would pass this leakage directly to the IF. Finally, in any case, the IF filters are necessary to ground the FETs' sources at the LO frequency. Keep the filters.

Because the RF connection is made to the FETs' drains, a virtual ground for both the IF and the LO, no filter is needed in the RF port. Avoiding an RF filter keeps the circuit simple. RF impedance transformation is rarely necessary; it's pretty easy to match the RF port without it.

A dc gate-bias circuit is not shown in Figure 7.6. The easiest way to provide gate bias is through large-value resistors connected from the dc bias source to the gates. Because the gates normally do not draw current, no dc voltage is dropped in this resistor. If the LO level is made too great, a gate resistor may help prevent excessive gate current caused by rectification in the gate-to-channel junctions, protecting the devices. The resistors also protect the FETs from electrical transients and static electricity, the two leading causes of dead FETs.

Finally, a note on layout. In most FETs the drain and the source theoretically are interchangeable when the dc drain-to-source voltage is zero. However, usually the source connection has considerably more metal and thus more parasitic capacitance than the drain. For this reason it is best to connect the lower-frequency port, the IF, to the FETs' sources.

7.3.3 Design

This circuit is really no different from the one in Section 7.2. Figure 7.7 shows the single-device equivalent circuit, which is obtained by splitting the mixer down the middle and transforming the source and load impedances appropriately (n is the turns ratio of the transformer).

Recognizing this equivalence, we can use the procedure in Section 7.2.3 to design the mixer. For simplicity, the initial design, steps 1 to 4 in the procedure on page 224, should be based on the single-device equivalent circuit. This circuit has the same conversion loss and port VSWR but only half the LO power of the complete balanced mixer.

7.3.4 Variations

Balanced Mixer Without an IF Balun

The circuit in Figure 7.8 has been suggested occasionally. It is arguably a true singly balanced mixer but has no IF balun. Still, it's not a good circuit, and it illustrates the problems we encounter when we try to sneak around the IF-balun requirement.

The series connection of the FETs at the IF frequency is the greatest problem with this circuit. Remember, we have stated frequently that the IF output impedance in any resistive mixer, diode or FET, is about 100Ω. This means that the output impedance of the two FETs in Figure 7.8 is about 200Ω, possibly higher. This may

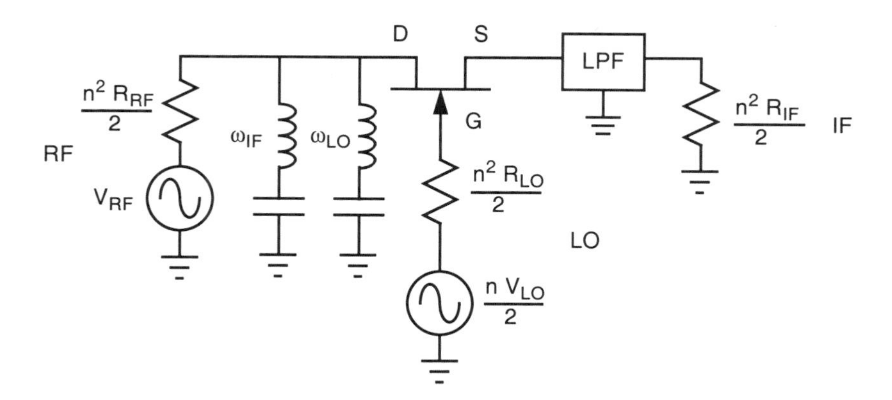

Figure 7.7 The single-device equivalent circuit of the singly balanced mixer shows that it is nearly identical to the single-FET mixer in Figure 7.4. Impedances and LO have been transformed to account for the separation of the circuit into two pieces; n is the turns ratio of the transformer at each port, and R_{LO}, R_{RF}, and R_{IF} are the port impedances. The resonators account for the LO and IF virtual ground at the RF port.

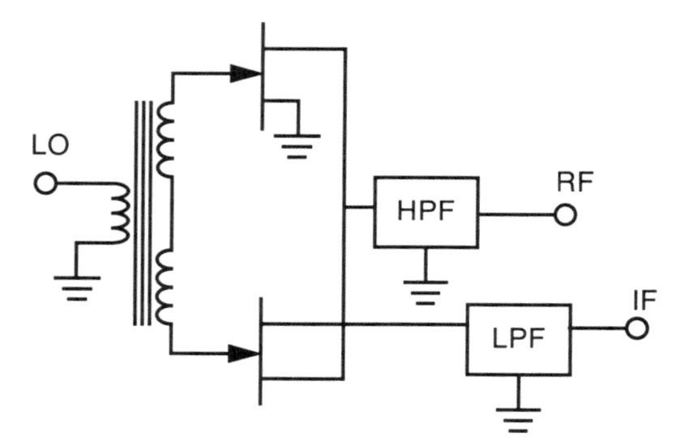

Figure 7.8 It is possible to remove the IF balun by putting the FETs' channels in series at the IF frequency. Unfortunately, this makes the IF output impedance quite high and difficult to match. It also introduces the need for an RF filter.

be difficult to match to a 50Ω IF load without the use of a transformer—just the structure we had hoped to eliminate!

In this circuit the RF connection to the FETs is no longer a virtual ground for the IF, and now an RF filter is necessary. This RF filter can be difficult to design; its input impedance, at the port connected to the FETs, must be very high at the IF frequency; otherwise, it will unbalance the mixer. We must retain the IF filter as well, and it is somewhat more critical. In the circuit in Figure 7.6, marginal IF filtering, whatever its other effects, does not unbalance the mixer. Here, however, the IF filter must ground the source of the lower FET very effectively at the RF and LO frequencies, or the mixer's balance will suffer, and there will be no virtual LO ground at the FET drains.

Clearly, this mixer is a tricky design, and making it work well is much more difficult than making the basic circuit work well. Still, the elimination of the IF balun may be worthwhile in some applications, especially ones where a high IF output impedance is acceptable. An example of the latter is a mixer followed by a low-frequency video IF amplifier.

Microwave Balanced Mixer

This circuit works well at high frequencies if the LO transformer is replaced by a microwave balun or a 180-degree hybrid, the sources are grounded, and the RF and IF both are filtered from the drains. If the IF frequency is not too high, the circuit can be relatively simple.

A possible configuration is shown in Figure 7.9. There a half-wavelength loop of transmission line is used for an LO balun. Although narrowband, this type of balun has an important advantage over most other types: it presents a short circuit to an

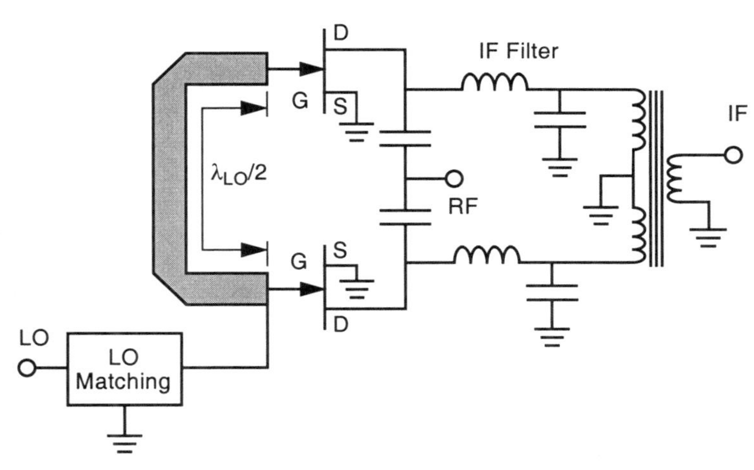

Figure 7.9 A singly balanced FET mixer appropriate for use at microwave frequencies. The capacitors in series with the RF port are for IF blocking.

even-mode excitation at its output ports. Thus, it effectively short-circuits RF leakage from the drain at the gate, the optimum termination. LO matching should be kept outside the balun; otherwise, the FETs must be tuned individually, unbalancing the mixer. In discrete circuits empirical LO tuning usually is adequate.

The high-frequency design has much higher levels of LO coupling through the gate-to-drain capacitance than the low-frequency design. Therefore, the virtual ground at the drain, which is inherent in this circuit, is all the more important. To prevent degradation of this property, the drains must be located very close together. The LO virtual ground is halfway between the drains, and any transmission line from the ground point to the drain is, in effect, a shorted stub. The LO-frequency terminating impedance at the drain, $Z_{d,\,LO}$, will then be

$$Z_{d,\,LO} = jZ_0 \tan(\theta) \tag{7.7}$$

where Z_0 is the characteristic impedance of the connecting line and θ is the distance from the drain to the short-circuit point, in degrees of phase at the LO frequency. If this line is, for example, one-quarter wavelength long, $\theta = 90$ degrees and the drain will see an open circuit, not a short. It is essential that the drains be connected to ground at dc. The tap in the IF transformer provides this connection nicely. The tap, by the way, is not essential for any other purpose; it can be replaced by an RFC or some other structure. Even a high-value resistor will provide the termination, as long as the LO level is kept low enough so dc gate current is not generated.

90-Degree Mixer

It is possible to make a 90-degree FET two-device resistive mixer. (You may notice a reluctance to call this a balanced mixer, because it is, at best, a very poor one.) The mixer requires 90-degree hybrids for applying the LO to the gates and the RF to the drains. It is possible to select the hybrid's ports for the RF and LO excitations in such a way that an IF balun is not needed. In this arrangement, however, the LO-RF isolation is poor and the drains are not virtual grounds for the LO. In the other possible configuration, a balun is needed, but the drains still are not LO virtual grounds and the mixer has all the disadvantages of a 90-degree diode mixer (Section 3.3). Forget this one.

7.3.5 Cautions

Probably because of the LO transformer's high inductance at low frequencies, this circuit seems to be unusually susceptible to the type of parasitic oscillation described in Section 7.2.5. The cure, as before, is to load the LO with extra resistance. The amount of resistance must be determined empirically, but usually it is only a few tens of ohms, at most.

7.4 DOUBLY BALANCED RING MIXER

Yes, FET resistive mixers come in doubly balanced forms, too. This is a very nice way to make a low-distortion RF mixer. Especially if you like transformers.

7.4.1 Characteristics

Because of the complexity of the ring and the interconnections between the FETs, this mixer is best suited for use at frequencies of, at most, a few gigahertz. Even in ICs, the interconnection parasitics and the unavoidable asymmetries hurt the performance of this circuit. Still, at frequencies up to about 18 GHz, careful layout may produce performance exceeding that of diode mixers.

This mixer has some very nice properties. The best is that all the FETs' drains and sources are virtual ground points for all undesired signals. The gates are all virtual grounds for the RF and IF. Because of these pleasant properties, the port-to-port isolation usually is very good, and this mixer can be used when the RF, LO, and IF bands overlap.

Finally, this is a true doubly balanced mixer, and it does a spectacular job of rejecting even-order IMD.

7.4.2 Description

Figure 7.10 shows the mixer. The FET ring operates as a commutating switch, much like the diodes in the ring and star diode mixers in Sections 3.4 and 3.6. As with the

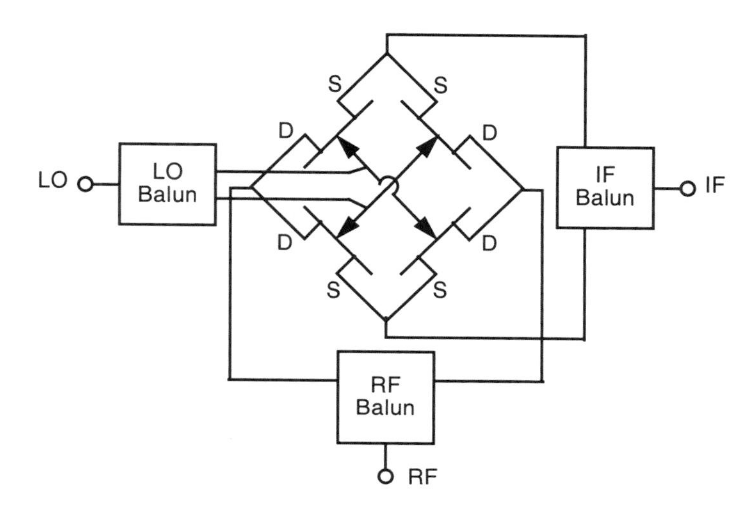

Figure 7.10 The doubly balanced resistive FET ring mixer is a commutating mixer, similar in concept to the diode ring and star mixers. The drain and source connections shown in the figure are not essential but usually work best.

diode mixers, it is easiest to understand this circuit if the FETs are viewed as switches. At any time, the two FETs on opposite sides of the ring are turned on, and the other two are off. When the LO voltage changes in polarity, the former FETs turn off and the formerly-off FETs turn on. This switching action connects the RF transformer to the IF transformer, reversing the polarity of the connection every half LO cycle. In effect, the RF is multiplied by a square wave.

Because of the symmetry of the circuit, the ring is, in effect, a balanced bridge at the RF and IF frequencies. Viewing the ring this way, we can see that there is no RF voltage between the nodes where the IF balun is connected and no IF voltage where the RF is connected. It is a little more difficult to see that the LO is coupled to each FET's drain and source nodes equally, so any pair of these nodes, on opposite sides of the ring, have zero LO voltage between them—the ideal situation!

In low-frequency mixers (and we did, after all, say that this is the most practical kind), wirewound transformers are used for the baluns. These provide some room for adjusting the impedances seen by the devices and also provide good balance. At higher frequencies, if you want to attempt such a design, an appropriate type of microwave balun can be used.

7.4.3 Design

Fundamental Design

Figure 7.11 shows the single-device equivalent circuit of the mixer. It is almost identical to the mixer in Figure 7.4, so the design approach in Section 7.2.3 is applicable.

The design of this mixer largely consists of adjusting the transformer turns ratios and the device size until reasonable VSWR has been achieved at all ports, the conversion loss is low, the LO power is within the required range, and the devices are as large as possible within these constraints.

These mixers are easiest to design when the equivalent circuit in Figure 7.11 is used for the initial design. With such a simple circuit, it is easy to perform the appropriate trade-offs, determine the performance, and determine the effect of nonzero-length interconnections. When the computer confirms that the circuit is working, the entire four-FET circuit can be simulated. If the single-device design is a good one, the complete circuit should work well with minimal effort.

DC Bias

As with all FET mixers, the simplest way to provide dc bias is through a center tap on the LO transformer and to ground the drains and sources through center taps on the RF and IF transformers. Unfortunately, many of the better types of transformers

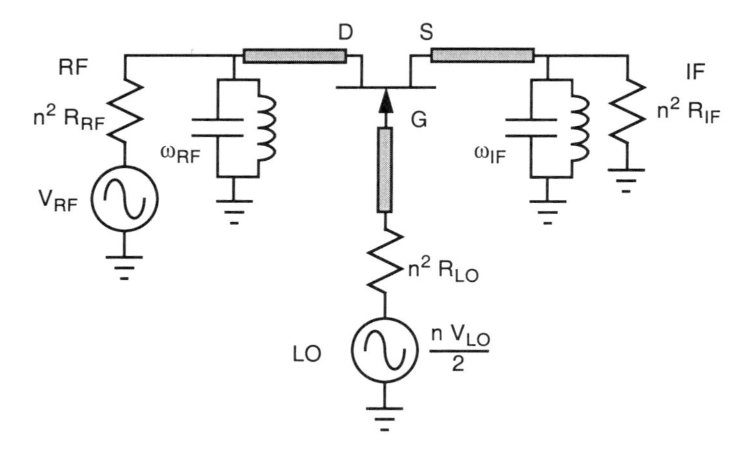

Figure 7.11 The single-device equivalent circuit of the mixer shows that it essentially is the same as the single-device mixer of Figure 7.4. The resonators are located at the virtual ground points, halfway between adjacent FETs' drains or sources. The shaded areas represent microstrip line connections between the devices. The turns ratio of the transformer at each port is n.

have no center taps, and frequently we want to use GaAs MESFETs but not negative dc bias.

Figure 7.12 shows one solution to the biasing problem. For negative bias, the dc voltage is applied to the gates as described above. For positive bias, the ring is "floated" at the dc bias voltage and the gates are grounded. Large-value resistors are used to couple the bias to the FETs, so no transformer taps or large inductors are needed. These are adequate as long as the LO level is kept low enough, as it should be, so no dc gate current exists in the FETs.

Layout

As frequency increases, the parasitics associated with all the interconnections become significant. Uncontrolled, interconnection parasitics can ruin the performance of the mixer. It is essential that the connections be as symmetrical as possible. The FETs should be connected drain to drain and source to source, and the lengths of the four interconnections between FETs should be equal. The node at which the IF and RF baluns are connected should be precisely halfway between each pair of drains or sources, and the LO connection should be precisely halfway between each pair of gates. The lengths of these connections should be only a few degrees of phase at the highest operating frequency. FETs' sources usually have greater stray capacitance than the drains; to minimize the effects of these parasitics, connect the IF to the sources and the RF to the drains, as shown in Figure 7.10.

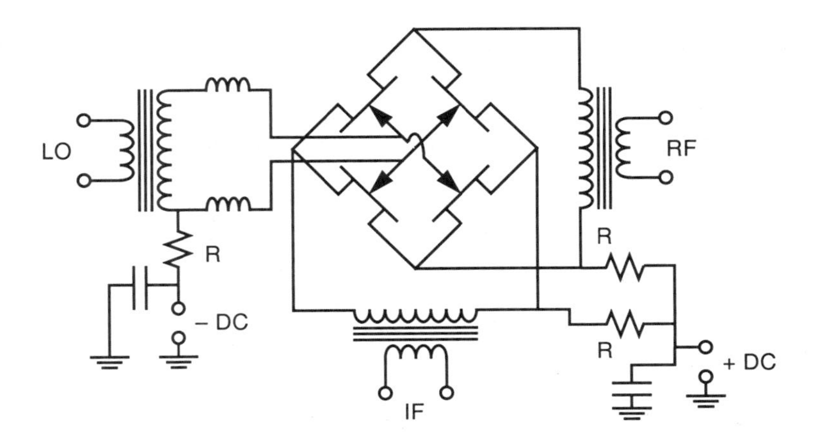

Figure 7.12 A complete mixer circuit. Transformers are used for the baluns, and the mixer can be biased with either positive or negative voltage. The unused bias port is simply shorted. The resistors *R* should be 1 - 2 kΩ, and the capacitors are for bypassing. Series inductors are used in the LO circuit to resonate the FETs' gate-to-channel capacitance.

Now that you know what to do, try to do it. It's impossible! When you try to lay out one of these mixers, you will see why some degree of asymmetry is simply unavoidable in a circuit this complex. Unless the frequency is very low, an IC is almost essential, at least for the FET "quad" if not the whole mixer, to minimize the asymmetries. The most difficult asymmetries seem to be the gate connections: it is almost impossible to connect the LO lines precisely halfway between the gates in any layout that equalizes the lengths of the drain and source interconnects. Each degree of phase difference results in loss of balance, and imbalance is manifest as limited even-order IMD rejection and port-to-port isolation.

7.4.4 Variations

The same trick we showed in Figure 7.8 to eliminate the IF balun in the singly balanced mixer can be used in this mixer. It has all the same problems, plus a few new ones. Now that the drains are no longer IF virtual grounds, we need RF filters, *two* of them, one in each lead from the RF transformer to the ring. We also need a pretty good IF filter. Don't attempt this variation unless you feel that you really understand this circuit. Even then, think about it.

7.4.5 Cautions

The most common problem in designing one of these mixers is achieving a good enough LO match to provide good IMD performance with reasonable LO power. In MESFET circuits it is easy to know whether the mixer is well matched and adequately pumped: you should see dc gate current when the gate bias is slightly increased or decreased. The instructions in *Tuning and Adjustment*, in Section 7.2.3, apply to ring FET mixers as well as to single-FET mixers.

7.5 SUBHARMONICALLY PUMPED FET RESISTIVE MIXER

There are at least two good reasons for using a subharmonically pumped mixer and one good reason not to. The reasons for using one are (1) a reduction in the complexity of the local oscillator and of the mixer itself, and (2) easy rejection of LO leakage into the RF and IF, especially at millimeter wavelengths, where the reactance of the FET's gate-to-drain capacitance is low. The reason for avoiding subharmonic mixers is sobering: virtually all aspects of their performance are worse than those of fundamental mixers.

7.5.1 Characteristics

In a subharmonically pumped mixer, we use two devices connected in parallel but pumped out of phase. This creates a conductance waveform for the pair that has only even harmonics of the LO frequency. Figure 7.13 shows the conductance waveforms

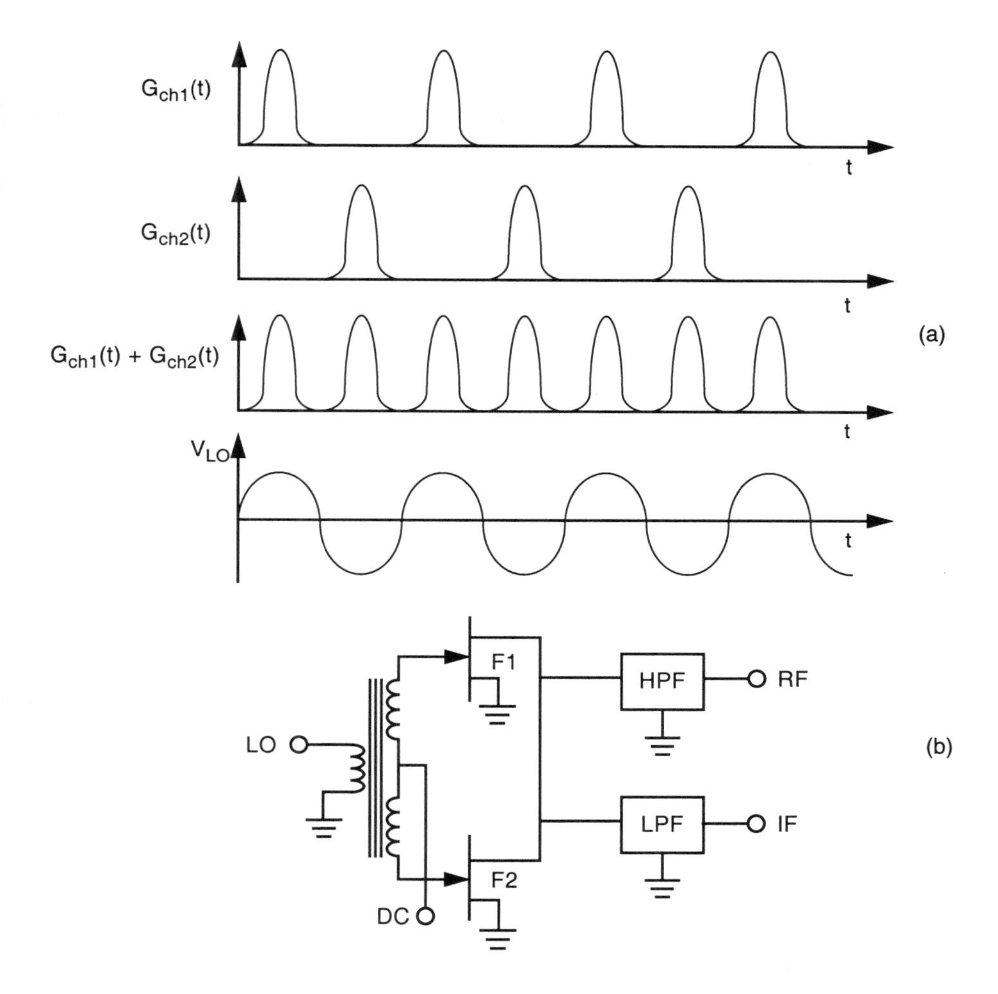

Figure 7.13 (a) Channel conductance waveforms in the FET subharmonically pumped mixer; (b) the mixer circuit.

and the basic circuit. The upper FET, marked F1, conducts when the voltage at its gate is positive; this is shown in the first curve of Figure 7.13(a). At the same time, the voltage at the gate of the other FET, F2, is negative, and if F2 is biased near pinch-off, the FET is off. The conductance waveform is shown in the second curve of Figure 7.13 (a). In the next half LO cycle, the situation is reversed; F1 is off and F2 is on. Because the FETs' channels are connected in parallel, the pair of FETs operates as a single device whose conductance is the sum of the channel conductances of the two FETs. This is shown in the third curve of Figure 7.13(a).

Basic Fourier-series theory shows that the conductance waveform of the combined FETs has no fundamental or odd-harmonic frequency components; it has only even-harmonic components. Since it has no fundamental frequency, there can be no mixing at the fundamental LO frequency, only at the LO's even harmonics. The strongest of these is, of course, the second; the fourth harmonic is much weaker, so the second-harmonic mixing is by far dominant.

As with the singly balanced mixer, the drain is a virtual ground at the fundamental LO frequency. It is not a virtual ground at the second LO harmonic, but this does not matter: the FET generates only very weak harmonics of the LO voltages and currents, primarily by the nonlinear gate-to-channel capacitance. For these reasons LO leakage into the RF and IF ports is very low and, at approximately half the RF frequency, is easy to filter out.

Similarly, AM LO noise, which is strongest close to the LO frequency, is not converted directly to the IF. Furthermore, as in all FET resistive mixers, the LO noise is not applied to the mixing element (the channel), so downconversion of AM LO noise usually is negligible.

The performance disadvantage of this mixer can be traced to the fact that the rise time of the conductance pulses is not as fast as it would be if the FET were pumped by a fundamental-frequency source, and their duty cycle is not the same either. Of course, the second-harmonic component of the conductance waveform of each FET always is weaker than the fundamental, so the conversion efficiency must be lower than in fundamental-LO mixers. The difference usually is at least 2–3 dB, giving the subharmonically pumped mixer at least 9–10 dB conversion loss. IM intercept points are also lower, but the difference depends on several factors, including operating frequency. Although it is difficult to generalize, a drop of at least 4–5 dB, compared to a fundamental, single-FET mixer, should be expected. The RF and IF port impedances also are harder to match and may be more sensitive to LO level. On the positive side, however, the half-frequency LO reduces the complexity of the LO source (and LO sources do have a tendency to become complex) and eliminates the IF hybrid, while retaining performance that is still good enough for many applications.

7.5.2 Description

Figure 7.13 shows about all there is to this circuit. The IF and RF circuits are, in effect, a diplexer; this is the most complex part of the circuit. The RF input and IF output impedances at each FET are relatively high; having two FETs in parallel helps to reduce these impedances, but they still may be difficult to match. The RF and IF filters shown in Figure 7.13 may have to include matching as well as filtering functions. Using relatively large devices also may help to reduce the impedances.

The circuit is appropriate for very high frequencies; subharmonically pumped FET mixers have been produced at millimeter wavelengths [6]. In such mixers, a microwave balun should be used for the LO; the simple half-wavelength balun and matching circuit shown in Figure 7.9 are just fine.

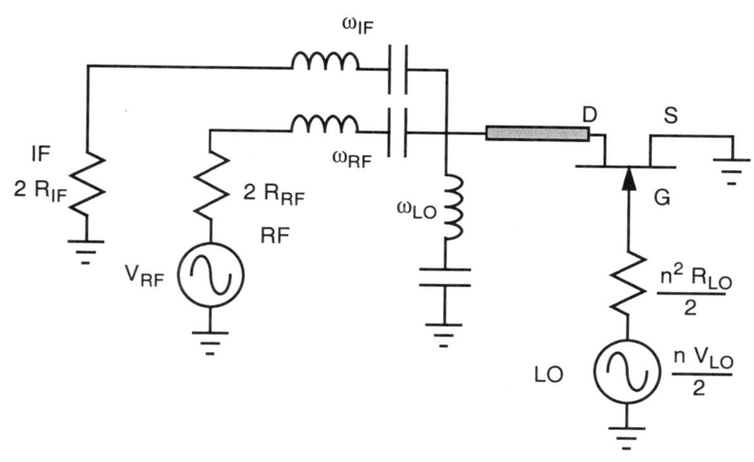

Figure 7.14 The single-device equivalent circuit is similar to those we have seen earlier. The ω_{LO} resonator accounts for the RF virtual ground at the drains, and the ω_{RF} and ω_{IF} resonators represent the diplexer. The piece of transmission line (shaded area) accounts for half the length of the connection between the drains.

7.5.3 Design

The single-device equivalent circuit of this mixer is shown in Figure 7.14. It is similar to those we already have seen and, as before, is a good place to begin the design. The process is the same as the one on page 224. Nonlinear analysis is essential here, because the circuit tends to be more sensitive to device size, LO level, and bias than a fundamentally pumped mixer.

The design of the gate matching circuit is identical to that of a singly balanced mixer, described in Section 7.3.3 for the low-frequency mixer and Section 7.3.4 for a microwave mixer. In the earlier mixers, we suggested adjusting the device size to help match the RF and IF ports, either by scaling the FET's gate width, in the case of an IC, or by selecting devices of different sizes, in a design for a hybrid circuit. That exercise is worthwhile here, but it may not be possible to achieve good port VSWR by adjusting device size alone.

7.5.4 Variations

For better or worse, there are not a lot of ways to make these. You're stuck with the one circuit.

7.5.5 Cautions

The magnitude of the second harmonic component of the conductance waveform, which is the dominant contributor to mixing, is much more sensitive to the shape of the conductance pulses than the fundamental-frequency component. As a result, the performance of this circuit usually is much more sensitive to LO level and dc gate bias than the fundamental mixer. This sensitivity alone may make a subharmonic mixer unsuitable for some applications.

As with other FET balanced mixers, be careful to minimize the length of the connection between the drains of the FETs. This should be no more than about 20 degrees long at the fundamental LO frequency. Similarly, be sure that the lines from the LO balun to the gates are equal in length. Remember that frequency multiplication is phase multiplication, so one degree of difference in these lines at the fundamental frequency is two degrees at the second harmonic.

REFERENCES

[1] Maas, S., "A GaAs MESFET Mixer with Very Low Intermodulation," *IEEE Trans. Microwave Theory Tech.*, Vol. MTT-35, April 1987, p. 425.

[2] Maas, S. A., "A GaAs MESFET Balanced Mixer With Very Low Intermodulation," *1987 IEEE MTT-S Int. Microwave Symp. Digest*, p. 895.

[3] Larson, L. E., *ed.*, *RF and Microwave Circuit Design for Wireless Communications*, Norwood, MA: Artech House, 1996.

[4] Geddes, J., P. Bauhahn, and S. Swirhun, "A Millimeter-Wave Passive FET Mixer with Low 1/*f* Noise," *1991 IEEE MTT-S Int. Microwave Symp. Digest*, p. 1045.

[5] Matthaei, G., L. Young, and E. Jones, *Microwave Filters, Impedance-Matching Networks, and Coupling Structures*, Norwood MA: Artech House, 1980.

[6] H. Zirath, I. Angelov, N. Rorsman, and C. Karlsson, "A W-Band Subharmonically Pumped Resistive Mixer Based on Pseudomorphic Heterostructure Field Effect Transistor Technology," *1993 IEEE MTT-S Int. Microwave Symp. Digest*, p. 341.

Chapter 8
Active Frequency Multipliers

The greatest disadvantage of diode frequency multipliers, discussed in Chapter 4, is that they use diodes. Because diodes are passive elements, diode frequency multipliers are lossy. Although varactor frequency multipliers theoretically are capable of 100% efficiency, in practice we are lucky to achieve a conversion efficiency of $1/n$ in a varactor multiplier, where n is the harmonic number. Resistive diode frequency multipliers are even worse. These have a theoretical optimum efficiency of $1/n^2$, but in practice 10-dB conversion loss is pretty good for a resistive doubler. The efficiency of resistive triplers is unspeakable.

Fortunately, FETs and bipolar transistors make dandy frequency multipliers. Their noise levels are decent (although, in this respect, nothing is as good as a varactor), their bandwidths are broad, and they can achieve conversion gain. These characteristics can simplify a frequency multiplier chain significantly.

Table 8.1 lists the circuits described in this chapter. The doublers are probably the most practical circuits. Triplers are difficult to produce successfully, and their performance is invariably poorer than the doublers'. It's a good idea to design the frequency-multiplier chain with this fact in mind.

8.1 ACTIVE FREQUENCY-MULTIPLIER THEORY

8.1.1 Why Use Active Multipliers?

In a single word, *efficiency*. Suppose we want to generate a 24-GHz signal by multiplying a 3-GHz signal, and we need to generate 10 dBm for a mixer local oscillator. Figure 8.1 shows two multiplier chains, one using diode multipliers and a second using FETs. RF and dc power levels are shown below the figures. The chain using passive multiplier stages needs several amplifiers to overcome the high losses of the multiplier stages, and these use a lot of power. The active multiplier chain, on the other hand, has approximately unity gain in each stage, so no amplifiers are necessary.

Table 8.1 Multiplier Circuits Described in This Chapter

Multiplier Type	Characteristics	Typical Applications
Single-device frequency multiplier	Good efficiency and bandwidth; fundamental-frequency leakage can be a problem.	General-purpose circuit; best used in ICs or other applications where small size is most important.
Balanced doubler	Very low fundamental-frequency and odd-harmonic leakage. Requires an input balun or hybrid. The optimum load impedance is easier to realize than with the single-device circuit.	This is the circuit to use for most doublers, unless you have a good reason not to.

Of course, we have "cooked the books" a little in our favor by assuming that the passive chain uses resistive multipliers, and these have greater loss than varactor multipliers. Still, a chain of varactor multipliers needs at least one or two amplifiers, and these still require more dc power than the active multiplier chain. Furthermore, cascaded varactor multiplier stages must have interstage isolation, and the cheapest isolator is often an amplifier. Finally, take a sober look at the warnings about the stability of varactor multipliers in Chapter 4. Do you really want to try to cascade three of those things?

In spite of its lower efficiency, a varactor frequency multiplier chain still may be appropriate in a few situations: (1) where LO noise—both phase noise and AM noise—are more important than efficiency, (2) where odd harmonics are needed, or (3) where very high harmonics must be generated from a low-frequency source (in this case, a step-recovery diode is best). Active multipliers have disadvantages, compared to varactors, in noise and in the generation of high or odd harmonics.

8.1.2 Active Multiplier Operation

Long ago, before transistors even existed, engineers designed frequency multipliers that used vacuum tubes. They quickly discovered what was necessary to make these beasts work well. The same principles, which we discuss here, apply to modern transistor frequency multipliers.

The Ideal Active Multiplier

An active frequency multiplier generates harmonics by rectifying the sinusoidal input signal. The principle is simple: the device—a FET or bipolar transistor—is biased near its turn-on point, and the input sinusoid turns the device on over part of

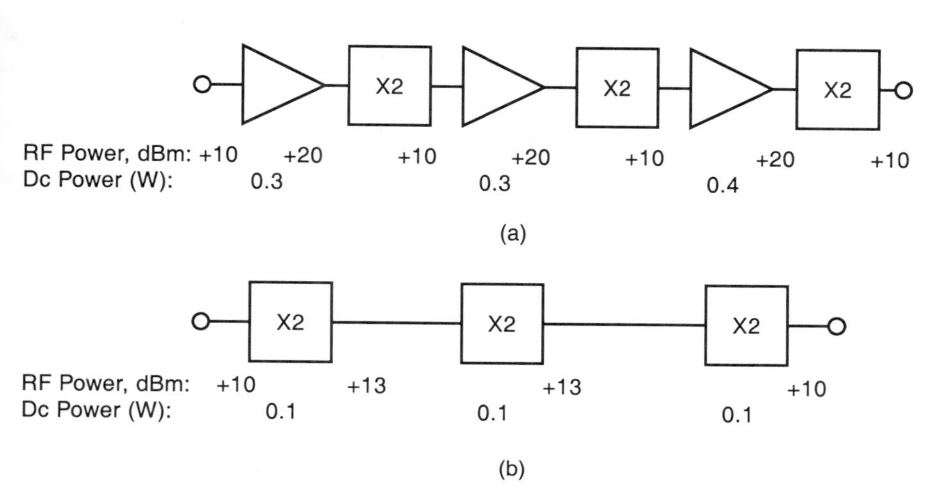

Figure 8.1 In general, a frequency-multiplier chain using passive elements (a) has lower dc-to-RF efficiency and more stages than a multiplier using active elements (b). The latter has greater efficiency because the active multiplier has approximately unity gain. The chain using passive multipliers must have intermediate amplifier stages, and because of the high loss of the multipliers, the amplifiers must have high gain and output power. In the example shown here, the passive-multiplier chain requires more than three times the dc power of the active-multiplier chain.

its cycle. The fraction of time that it is on, compared to the entire period, called its *duty cycle*, is adjusted to maximize the desired output harmonic. The higher the harmonic, the shorter the duty cycle must be. For a doubler, the optimum is around 30%; for a tripler, about 20%.

Figure 8.2 shows a somewhat idealized active FET multiplier. (To retain some degree of concreteness in the following discussion, we describe only a FET frequency multiplier. The principles, however, are applicable to any active device.) The gate is driven by an RF source, V_s, at the input frequency, ω_1, and gate bias, V_{gg}, is applied as well. We assume that capacitive parasitics at the drain and gate are absorbed into the tuned circuits and, for now, simply neglect parasitic resistances. Another LC circuit at the drain is tuned to a harmonic ω_n of the excitation frequency, $\omega_n = n\,\omega_1$. This tuned circuit allows only nth-harmonic components of drain voltage across the load, R_L, and also across the channel. The dc drain voltage is V_{dd}.

The dc drain bias is the same value used for the FET in such ordinary applications as small-signal amplification. The gate bias voltage is somewhat below (more negative than) pinch-off. Figure 8.3 shows the resulting waveforms in the multiplier; the FET conducts in pulses that occur whenever the gate voltage is above the pinch-off voltage, V_p.

The duty cycle of the drain-current pulses can be adjusted by changing the dc gate bias, V_{gg}. Clearly, for best efficiency, we want the peak drain current to be as

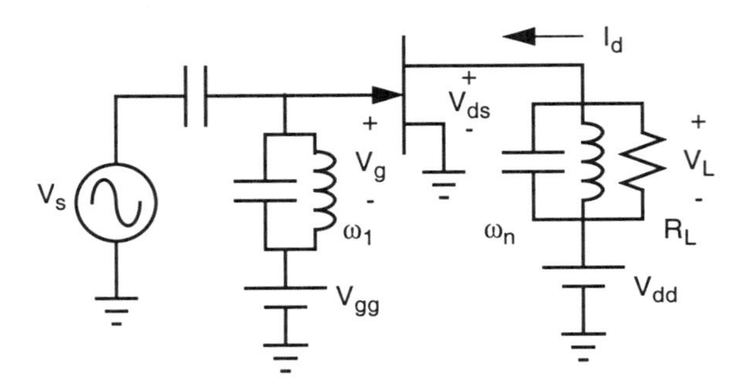

Figure 8.2 A somewhat idealized representation of a FET frequency multiplier. We assume that the FET's drain and gate parasitic capacitances have been absorbed into the tuned circuits and, in the accompanying discussion, that resistive parasitics are negligible.

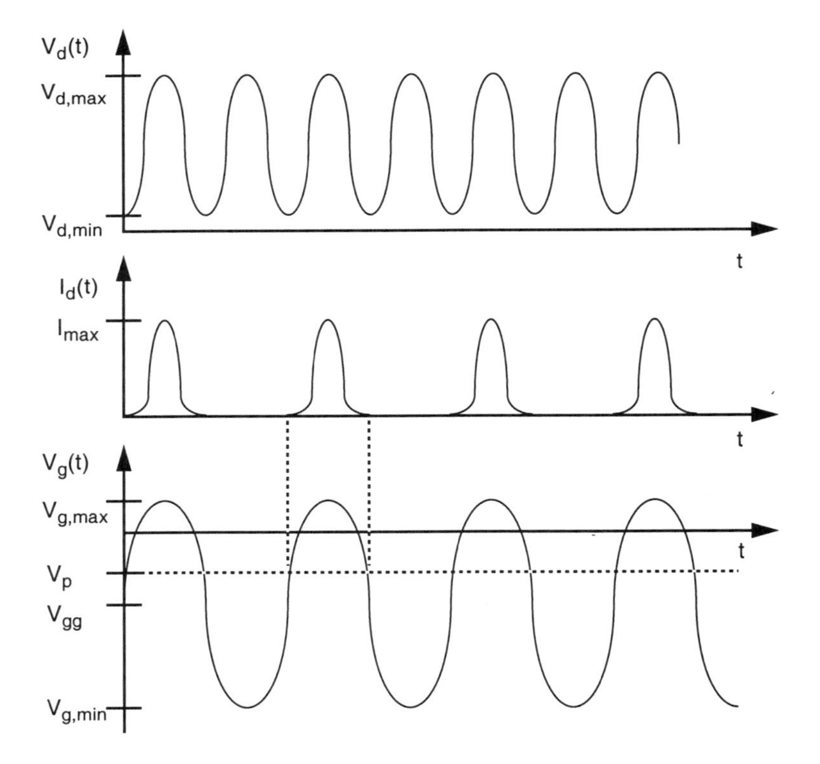

Figure 8.3 Voltage and current waveforms in the FET frequency multiplier. The FET's channel conducts in pulses when $V_g(t) > V_p$. These pulses generate a second-harmonic voltage at the output.

great as possible, I_{max}, so $V_g(t)$ must peak at $V_{g,\max}$; as we reduce the gate bias, to reduce the duty cycle of the pulse, we also must adjust the level of the input excitation to compensate. A shorter duty cycle, which is necessary for higher harmonics, requires more negative V_{gg} and greater RF input voltage (and, therefore, more RF power). As we continue to decrease the duty cycle in this manner, eventually the negative peaks of the gate voltage reach the gate-breakdown voltage, and we can reduce the duty cycle no further.[1]

The drain-current pulses can be modeled as a train of rectified cosine pulses. The time axis is arbitrary, so the pulses can be centered at $t = 0$ to remove the phase terms in the Fourier series. The drain current, then, is

$$I_d(t) = I_0 + I_1 \cos(\omega_1 t) + I_2 \cos(\omega_2 t) + \ldots \quad (8.1)$$

where I_n is the nth-harmonic current component, ω_n, given by

$$\begin{aligned} I_n &= I_{max} \frac{4t_0}{\pi T} \left| \frac{\cos(n\pi t_0/T)}{1-(2nt_0/T)^2} \right| & n > 0 \\ &= I_{max} \frac{2t_0}{\pi T} & n = 0 \end{aligned} \quad (8.2)$$

and I_{max} is the maximum channel current, t_0 is the length of the pulse, and T is the period of the fundamental frequency. When $n > 0$ and $t_0 / T = 0.5 / n$, (8.2) has some obvious problems. In that case,

$$I_n = I_{max} \frac{t_0}{T} \quad (8.3)$$

From Figure 8.3 we see that the peak RF drain voltage can be no more than V_{dd}, regardless of I_n. To maximize output power at the nth harmonic, we must maximize I_n; it's the only option. To maximize I_n we know that we must keep the peak value of I_d constant at I_{max}; (8.2) shows that the only remaining degree of freedom is the duty cycle, t_0 / T. This is what we need to optimize.

Figure 8.4 shows I_n / I_{max} as a function of the duty cycle. There are distinct optima for all values of n. For $n = 2$, the optimum is $t_0 / T = 0.32$; for $n = 3$, it is $t_0 / T = 0.21$. The latter value is unachievable in most MESFETs, as is the optimum for $n = 4$, approximately 0.18. Interestingly, the next peak in the $n = 4$ curve is at $t_0 / T = 0.50$, a much more practical value. Although the output power is a little lower than $t_0 / T \sim 0.30$, this power is achieved with lower input level and, therefore, better conversion gain. Consequently, $t_0 / T = 0.50$ probably is no worse, in practice,

---

1. This is really a lie. There is a way to reduce the duty cycle further: allow the peak gate voltage to decrease as V_{gg} decreases and, consequently, the peak drain current. This usually results in lower harmonic output power than simply using a nonoptimum duty cycle. It's a Pyrrhic victory.

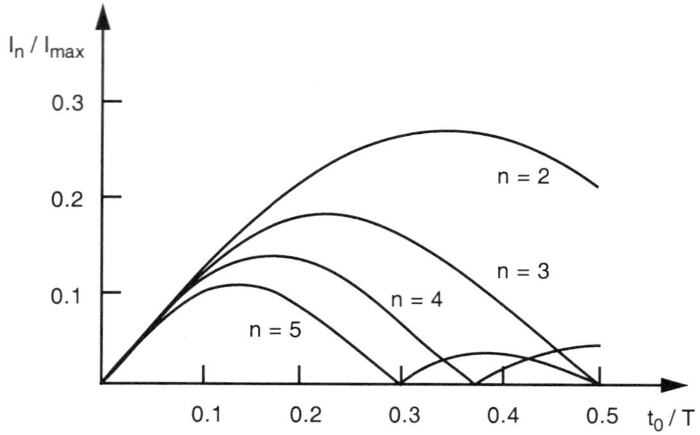

Figure 8.4 Magnitudes of the harmonic current components of $I_d(t)$, normalized to the maximum channel current, I_{max}.

than $t_0 / T \sim 0.30$. For a fourth-harmonic multiplier, avoid $t_0 / T \sim 0.38$!

We need an expression for the duty cycle as a function of the voltages in the gate circuit; it is

$$\theta_t = 2\,\text{acos}\left(\frac{2V_p - V_{g,\,max} - V_{g,\,min}}{V_{g,\,max} - V_{g,\,min}}\right) \tag{8.4}$$

where θ_t is the conduction angle, $2\pi t_0 / T$. The gate-bias voltage that achieves this value of θ_t is

$$V_{gg} = \frac{V_{g,\,max} + V_{g,\,min}}{2} \tag{8.5}$$

and the peak RF gate voltage is

$$|V_g(t)| = V_{g,\,max} - V_{gg} \tag{8.6}$$

The rest of the design is all downhill. The maximum peak-to-peak RF drain voltage, at the nth harmonic, is 0.5 $(V_{d,max} - V_{d,min})$. When the output power at the nth harmonic is optimized, the peak RF voltage is

$$|V_L(t)| = I_n R_L = \frac{V_{d,\,max} - V_{d,\,min}}{2} \tag{8.7}$$

here $V_{d,max}$ and $V_{d,min}$ are defined in Figure 8.3. R_L can be found from (8.7). The output power at the nth harmonic is

$$P_{L,n} = \frac{1}{2}I_n^2 R_L = \frac{1}{4}I_n(V_{d,max} - V_{d,min}) \tag{8.8}$$

and the dc power is

$$P_{dc} = I_0 V_{dd} = \frac{2t_0}{\pi T} I_{max} V_{dd} \tag{8.9}$$

The dc-to-RF efficiency is just the ratio of (8.9) to (8.8).

A multiplier's most important performance parameter is its conversion gain. In an active multiplier the gain can be greater than unity. We already have an expression for the output power, (8.8); we need an expression for the available input power in terms of the RF gate voltage. If we simplify the input circuit to a series combination of resistors and the gate-to-source capacitor, C_{gs}, the input power can be found easily. When the input is conjugate matched, the input power equals the available power from the source, P_{av}. Therefore,

$$P_{av} = \frac{1}{2}(V_{g,max} - V_{gg})^2(R_s + R_i + R_g)(\omega_1 C_{gs})^2 \tag{8.10}$$

where R_s, R_i, and R_g are the source, intrinsic, and gate resistances of the FET, respectively. The ratio of (8.10) to (8.8) is the conversion gain.

Feedback Effects

A short-circuit drain at the fundamental frequency usually results in the best overall performance but not necessarily the best performance in any one respect. In triplers and higher-harmonic multipliers, for example, a fundamental-frequency drain short results in disappointingly low conversion gain. By adjusting the fundamental-frequency termination, we can achieve much better gain but at the expense of reduced stability margin. This trick is somewhat like using positive feedback in an amplifier; it is a little dangerous. Still, it may be worthwhile if the alternative is an unacceptably low conversion gain.

An illuminating description of feedback effects in a FET multiplier can be found in a classic and very readable paper by Rauscher [1]. Figure 8.5 is a reproduction of one of his figures showing the dependence of conversion gain on the fundamental-frequency drain termination. The latter is described by θ, the length of a hypothetical 50Ω transmission line between the FET's drain terminal and the short-circuit reference plane of the filter:

$$X(\omega_1) = 50\tan(\theta) \tag{8.11}$$

It is interesting to note that the gain becomes infinite, signifying oscillation, at a moderate inductive reactance. This corresponds to resonating the FET's drain-to-source capacitance. The resonance maximizes the fundamental-frequency drain-to-source voltage and causes high fundamental-frequency feedback. At $\theta = 90$, corresponding to an open-circuit termination, the gain is very low, and at $\theta = 0$ and $\theta = 180$, the short-circuit case, the gain is somewhere between these extremes.

Interestingly and in contrast to the high sensitivity of gain to fundamental-frequency termination, the maximum output power is largely invariant with this termination. Feedback affects gain, not output power.

This result shows that it may be worthwhile to use a little fundamental-frequency feedback to increase gain in high-order multipliers, whose gain is characteristically low. In triplers there often is a reasonably broad region of load reactances where gain is improved but stability margin does not suffer unduly.

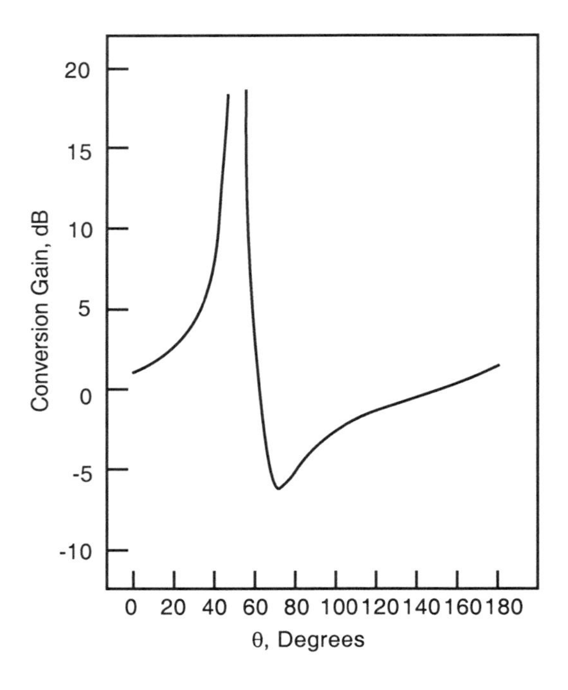

Figure 8.5 Effect of fundamental-frequency drain terminating impedance on conversion gain of a doubler. See (8.11) for the meaning of θ. From [1], © 1983 IEEE; reprinted with permission.

8.2 SINGLE-DEVICE MICROWAVE FREQUENCY MULTIPLIER

8.2.1 Characteristics

The most important characteristic of a FET frequency multiplier, as we already have noted, is the possibility of obtaining conversion gain instead of loss. The gain depends on a number of factors, and a few minutes experimenting with (8.2) through (8.10) should show what possibilities exist for achieving the desired gain and output power. Generally, ordinary small-signal MESFETs can achieve conversion gain in doublers at frequencies up to at least 16 GHz, possibly higher. Because of their smaller parasitics, especially C_{gs}, HEMTs can achieve gain up to considerably higher frequencies than MESFETs, often well into the millimeter range. However, HEMTs have relatively low values of I_{max}, so output power may not be as high as with MESFETs at lower frequencies.

Because I_{max} is proportional to gate width, output power depends largely on the size of the device. An output power of 10 dBm from a single small-signal MESFET at 10 GHz, with a decibel or two of gain, is typical. There is indeed a trade-off of power against gain. To achieve gain at higher output frequencies, C_{gs} must be reduced, and this requires a narrow device. A narrow device results in lower output power.

FET and BJT frequency multipliers can be made very broadband, but again there is a price: increasing the bandwidth reduces conversion gain. As with other FET circuits, the FET's input Q may be very high, and if the multiplier's bandwidth is more than a few percent, the input may be impossible to match over the whole band. The equation for available input power, (8.10), is based on the assumption that the input is matched. If it is not, reflection losses will reduce the gain to a value lower than predicted by (8.8) and (8.10).

One of the most important considerations, as in any multiplier, is fundamental-frequency leakage at the output. The FET is, after all, an amplifying device, and it amplifies the applied fundamental-frequency excitation. Thus, it is more difficult to filter the output of an active multiplier than a passive one. Balanced multipliers, described in Section 8.3, are a nice solution to this problem.

Doublers are the most practical active multipliers. Triplers and higher order frequency multipliers have low conversion gain, perhaps no better than a resistive diode doubler. When high harmonics are needed, it is best to avoid the use of high-harmonic stages. The ideal situation is to use a chain of doublers or, at worst, a tripler and some doublers. Try to avoid using quintuplers or higher-harmonic multipliers in frequency-multiplier chains, or be prepared to amplify their outputs and to use a lot of filtering. Note that the efficiency of these circuits depends largely on the fundamental nature of the multiplication process and applies to all types of devices. Using a different device will not avoid the inherent inefficiency of high-harmonic multiplication. The universe was created this way.

8.2.2 Description

Figure 8.6 shows a single-device FET frequency multiplier circuit appropriate for microwave use. The input matching circuit uses a single inductor and a quarter-wave transformer, and the output circuit uses a filter consisting of a series of high- and low-impedance sections, all one-quarter wavelength long. The bias circuits use a high-impedance line and a low-impedance, open-circuit stub.

The output filter is relatively large, but necessarily so. Although a simple quarter-wave open-circuit stub, connected to the FET's drain, might be a seductive alternative, a single stub usually does not have a high enough Q to provide good fundamental-frequency rejection. It is also rather narrowband.

We use our earlier approach, described in Section 6.2.3, to design the input matching circuit.

8.2.3 Design

Equations (8.4) through (8.10) are all we need to perform an initial design of a multiplier. Once the initial design is in place, it can be optimized by harmonic-balance analysis. The initial design will be surprisingly close to the final one; the greatest difference probably will be in the input matching. The design consists of an initial estimate of the load impedances and resulting gain and efficiency. Knowing these, we can design the input and output matching circuits. The initial design finally is optimized on the computer by means of nonlinear analysis.

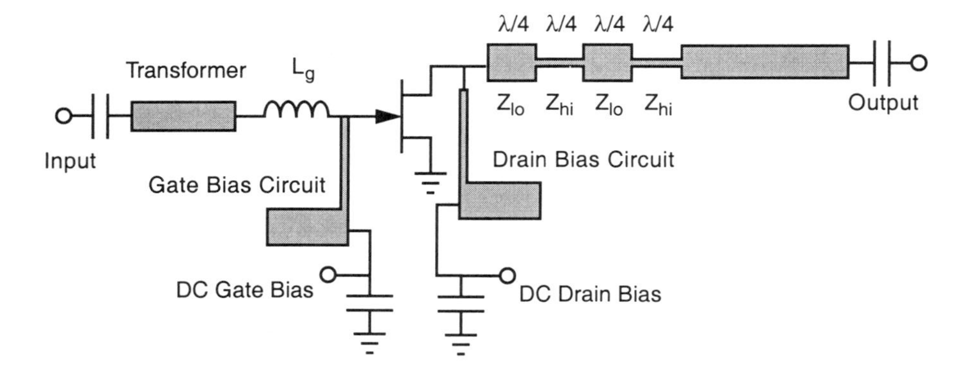

Figure 8.6 A microwave FET frequency doubler. The input matching and output filter are similar to those used in FET mixers (Figure 6.3).

Device Selection

As we shall see when we apply some numbers to (8.4) through (8.10), a conventional small-signal MESFET makes a pretty good frequency multiplier. These devices can achieve a few decibels of gain at frequencies below about 12 GHz. If higher power is needed and the frequency is not too high, a power device can be used. At higher frequencies, the input capacitance of a power device may limit the gain too severely.

Initial Design

First determine the basic operating parameters of the MESFET:

1. Select an optimum value of t_0 / T from Figure 8.4 and note the corresponding value of I_n / I_{max}.
2. Use (8.4) and (8.5) to determine $V_{g,min}$ and V_{gg}.
3. Use (8.10) to determine the available input power. Remember, this is for a conjugate match.
4. Determine a reasonable value of dc drain bias V_{dd}. *Do not make this value too great*; remember that the peak reverse gate voltage is very large, and the drain-voltage swing will be large as well.
5. Estimate values of $V_{d,max}$ and $V_{d,min}$. For a small-signal MESFET, $V_{d,min}$ should be about 1V and $V_{d,max} = 2\ V_{dd} - V_{d,min}$.
6. Estimate the load resistance, R_L, from (8.7) and the output power from (8.8).
7. The available input power and output power are now known, and from these the gain can be calculated. If the multiplier is broadband, it may be necessary to reduce the anticipated gain by a decibels or so to account for input reflection loss.
8. Finally, estimate the input impedance as

$$Z_{in} = R_s + R_i + R_g + \frac{1}{j\omega_1 C_{gs}} \tag{8.12}$$

It is interesting to apply some numbers to these equations. For example, let's look at R_L. An ordinary small-signal MESFET has $I_{dss} \cong 100$ mA, so for a doubler I_{max} is a little above this value and $I_n \cong 30$ mA. If we select $V_{dd} = 3.0$V and $|V_L(t)| = 2$V, we obtain $R_L = 67\Omega$, a close match to our ideal of 50Ω. A 50Ω load would give $|V_L(t)| = 1.5$V and an output power of 13.5 dBm. This performance is not bad at all, so it might be best simply to run our FET's second-harmonic output through the drain filter and into a 50Ω load. Of course, we must do something about the FET's parasitic drain-to-source capacitance, but a judicious tweak of the output filter, on the computer, probably will take care of it.

The design process is as follows:

1. Design the input matching circuit. Use the input impedance from (8.12). For moderate bandwidths, less than approximately 30% of the center frequency, the method described in Section 6.2.3 can be used to design a matching circuit. That method involves resonating the input capacitance with a series inductor, $L_g = 1 / \omega_1^2 C_{gs}$. The transformer in Figure 8.6 is designed to change the 50Ω source impedance to a value of resistance, R, that provides the desired bandwidth, $\Delta\omega$:

$$\Delta\omega = \frac{R + R_s + R_i + R_g}{\omega_1 L_g} \tag{8.13}$$

   This results in an input mismatch, which decreases gain. For more broadband matching circuits, it is necessary to synthesize a network. See [2].

2. Design the output filter and matching circuit. If the operating parameters and device size have been adjusted so that the ideal load is 50Ω, the design is simple. The output filter consists of a cascade of low- and high-impedance sections. The impedances of these sections should be as low and high as the technology allows; in microstrip on alumina, for example, the values will be about 25Ω and 100Ω, respectively. The sections are one-quarter wavelength long at the fundamental frequency; at the second harmonic, they are half-wavelength, and thus electronically invisible. Four sections are usually enough for moderate bandwidths.
3. If a value of R_L other than 50Ω is needed, follow this section with a transformer. Because the filter is an integral number of half-wavelengths, output of the filter is electrically equivalent to the FET's drain terminal.
4. The high-impedance line in the drain dc bias circuit can be made a little shorter than one-quarter wavelength to make it inductive. This inductance resonates the drain-to-source capacitance at the output frequency.
5. For other approaches to the design of this filter, see the sections on half-wave filters in Matthaei [2].
6. Optimize this structure on the computer. If you do not want to use the drain-bias circuit to compensate for the FET's drain-to-source capacitance, C_{ds}, you may be able to tweak this structure to do so. (Try both methods to see which works better.) Simply terminate the FET end of the structure with a load consisting of R_L and C_{ds} in parallel, and optimize for (1) a conjugate match at the second harmonic and (2) a short circuit at the FET end at the fundamental frequency. Include the drain dc bias circuit in the analysis.
7. Finally, design the bias circuits.
8. Optimize the entire circuit on the computer by harmonic-balance analysis. Treat the input level and gate bias as tuning adjustments and adjust the gate inductance, L_g, and transformer for an input VSWR that is as good as

possible over the required bandwidth. For uniform conversion gain over the band, you may need to match the input best at the high-frequency end of the band.

9. The drain circuit should require very little adjustment at this point. It is best not to adjust the drain filter; once you have it right, don't mess with it! If you adjust this filter, you may indeed improve the gain and output power, but you also may destroy the fundamental-frequency short circuit at the drain. The imperfect termination may make the multiplier unstable and cause high fundamental-frequency leakage. To adjust the harmonic load impedance, make changes on the output side of the tuner. To adjust the fundamental-frequency drain termination, use a length of line between the FET's drain and the filter and *be careful*!

8.2.4 Variations

At lower frequencies, MOSFETs and JFETs work well as frequency multipliers. The design approach is essentially the same as the one in Section 8.2.3, except for obvious differences related to the differences in the devices. At frequencies up to at least 1 or 2 GHz, carefully designed lumped-element matching circuits can be used, and simple parallel L-C resonators usually have high enough Q to provide adequate rejection of unwanted harmonics.

BJT frequency multipliers also are practical at frequencies up to about 1–2 GHz. BJTs are often less stable than FETs in circuits that are strongly driven. There are at least two reasons for this behavior: first, the low-frequency gain of a BJT is much higher, and this is conducive to low-frequency parasitic oscillations. These modulate the multiplier's output power and appear on the spectrum analyzer as a broad, triangular spectrum of discrete tones.[2] Second, the nonlinearity of the base-to-emitter capacitance in BJTs is much stronger than the nonlinearity of the gate-to-source capacitance in FETs. This reactive nonlinearity also contributes to strange parasitic oscillations when the BJT is strongly driven.

To prevent BJT instability, be sure that the connections to the device are short and have no more inductance than necessary. Also, make certain that the emitter is well-grounded or bypassed and that the base is terminated in a low resistance, a few ohms, at low frequencies. This base-terminating resistance can be part of the bias network.

Both BJTs and FETs can be self biased. To do this, the base or gate is kept at dc ground potential and a resistor is connected in series with the emitter or source. The resistor's value, R_b, is

2. The technical term for this behavior is *Christmas treeing*, a term that could be created only by engineers.

$$R_b = \frac{I_0}{V_b} \tag{8.14}$$

where I_0 is the dc collector or drain current and V_b is the gate or base bias voltage. I_0 is given by (8.2).

The self-bias resistor must be bypassed at the operating frequencies. The bypass capacitor must have a low impedance not only at the fundamental and output frequencies but at all other harmonics up to at least the third. Remember that chip capacitors have a series self-resonant frequency, above which the chip capacitor is inductive. The manufacturer should be able to provide the self-resonant frequency of the capacitor. Most chip and packaged FETs and BJTs have two source contacts or leads. Do not connect both source contacts together and use a single chip capacitor; use two capacitors instead, one for each contact.

When self bias is used, the gate or base bias voltage varies with drive level. This behavior is actually an advantage: variable bias tends to compensate for overdrive. It provides a more constant output power with variations in the drive level than does fixed bias.

Dual-gate FETs have been suggested as frequency multipliers. Their theory is somewhat more complex and their operation somewhat less intuitive. These are a bit beyond the scope of this book, but we provide some references [3–5] for anyone interested.

8.2.5 Cautions

The most important cautions already have been stated, but they are important enough to bear repeating: (1) the efficiency of high-harmonic multipliers may be poor; (2) stability, especially in BJT multipliers, requires vigilance; and (3) careful filtering is needed to provide a good drain short at unwanted harmonics and to minimize the fundamental-frequency output.

8.3 BALANCED FREQUENCY DOUBLER

8.3.1 Characteristics

This is probably the most practical way to make an active frequency doubler. It should be considered the default approach; the single-device circuit should be used only when the balanced circuit is not acceptable. The most likely reasons for rejecting the balanced circuit, in favor of the single-ended one, are its larger size and greater dc power.

This circuit uses two FETs. It has an input balun or a 180-degree hybrid but dispenses with the output filter. Additionally, when small-signal FETs are used, the circuit provides a good direct match to a 50Ω load, so the output circuit can be very simple indeed.

The greatest advantage of this circuit is its very good fundamental-frequency and odd-harmonic rejection. The degree of rejection depends on the balance of the input hybrid, but 20-dB additional rejection at microwave frequencies is typical. At lower frequencies, where good balance is easier to maintain, greater rejection is possible.

As with other balanced circuits, this multiplier provides twice the output power of a comparable single-device circuit. The gain is approximately the same, so twice the input power must be provided as well.

This circuit is most appropriate for doublers. Clearly, it can be used only for even-harmonic multipliers. In theory it could be used for a quadrupler, but such a circuit would have very low efficiency and would require filtering—just what we want to avoid—to reject the second harmonic. In most cases it is better to cascade two of these circuits than to make a single-stage quadrupler.

8.3.2 Description

Figure 8.7 shows the basic circuit. The gates are driven out-of-phase at the fundamental frequency, and the drains are connected in parallel. As with the single-device circuit, the FETs are biased below pinch-off, so each conducts over somewhat less than one-half the LO cycle.

Since the FETs are driven out of phase, one FET conducts on the positive LO peaks and the other on the negative. With the drains in parallel, the result is a doubling of the drain-current pulses. The resulting pulse train has only even harmonics; it contains no fundamental frequency or odd harmonics, so there can be no output at those frequencies.

This circuit has a virtual ground at the drains at the fundamental frequency. Thus, the drains are optimally terminated at the fundamental frequency, simply because of the symmetry of the circuit. No filter is needed to provide the optimum termination.

Figure 8.8 shows a microwave realization of the circuit. A rat-race hybrid provides the phase split. The connections from the hybrid to the gates must be equal

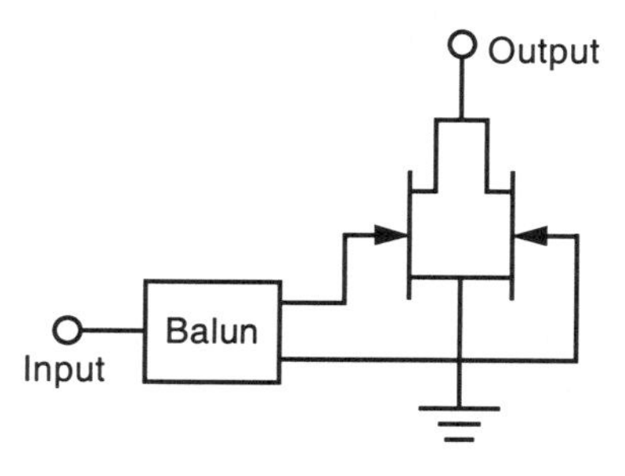

Figure 8.7 Basic circuit of the balanced multiplier. Either a balun or a 180-degree hybrid can be used.

in length and include a quarter-wave transformer to match the gates. An inductor, L_g, resonates the FET's input capacitance. A quarter-wave transformer at the output provides the optimum load impedance to the FETs; since the optimum load often is very close to 50Ω, this transformer may be unnecessary. The output capacitances of the FETs can best be resonated by making the high-impedance part of the dc bias line a little shorter than one-quarter wavelength. The dc bias circuits at the gates and drains are conventional.

8.3.3 Design

The circuit in Figure 8.8 is essentially two single-device frequency multipliers combined at their inputs by a hybrid and connected in parallel at their outputs. Because of the parallel connection, each multiplier sees an impedance equal to twice the load impedance of the pair. This has a distinct advantage: with smaller devices or HEMTs, which would be used in a high-frequency doubler, the balanced circuit's optimum load is nearly 50Ω.

Recognizing this point, we see that the design of a balanced multiplier is essentially the same as the design of a single-device multiplier. We change only the load impedance. Here is the process:

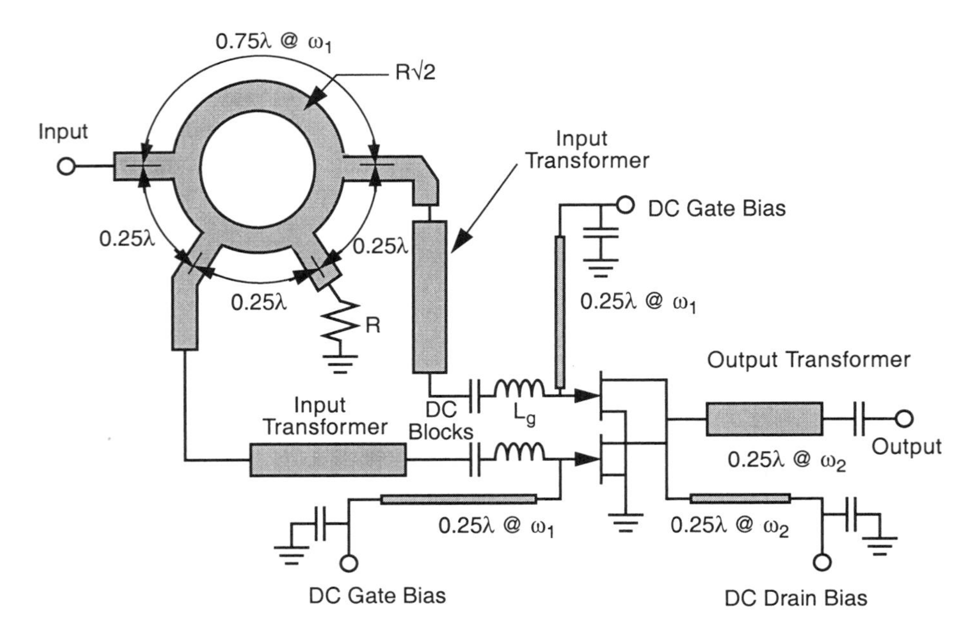

Figure 8.8 A balanced FET frequency doubler using a hybrid to provide the input phase split. The connections between the hybrid and the FETs must be equal in length.

1. Design a single-device multiplier according to the process described in Section 8.2.3. Remember that the load impedance, R_L, in this process is twice the load impedance that will be used for the complete balanced multiplier.
2. As in the single-device circuit, the optimum load impedance is a resistance in parallel with an inductive reactance (which resonates the FETs' drain-to-source capacitances). The length of the drain-bias line can be made shorter than one-quarter wavelength to serve nicely as the reactance. (Remember, this reactance must be half that determined from the single-device circuit, because it resonates two capacitances.) If the ideal load impedance differs significantly from 50Ω, a quarter-wave output transformer can be used to present the proper load impedance to the FET pair.

8.3.4 Variations

A Gilbert mixer or a similar FET doubly balanced mixer circuit (Sections 6.6 and 6.7) often can be used as a balanced frequency doubler. Just connect the outputs in parallel, with appropriate matching, instead of using an output balun. Even though four devices are used in the circuit, it is really a pair of singly balanced multipliers connected in parallel, not a doubly balanced multiplier. There is really no such thing as a doubly balanced active frequency multiplier.

A complete rat-race hybrid is often unnecessary. A 180-degree microstrip line can be used as a simple balun. Such a balun has a bandwidth of approximately 15%. See Figure 8.9 for an example of this type of circuit.

For broader bandwidth, another type of balun can be used. One option is to use a Marchand balun similar to the ones described in Section 3.6.2. Since the output circuit of a balanced multiplier inherently is very broadband, a broad bandwidth is easy to obtain. Unfortunately, few good 180-degree hybrids exist; the one shown in Figure 3.4 is about as good as they come. At low frequencies a transformer balun can be used.

The bias circuits are conventional. Figures 8.8 and 8.9 show bypass capacitors at the gate- and drain-bias terminals, but an open-circuit, low-impedance stub can also be included. It's a good idea to use a large (~1,000Ω) resistor in series with the gate-bias terminals to protect the FETs and to help decouple the devices from the bias supply. Both circuits show two separate gate-bias terminals, but usually these are connected in parallel to a common dc bias source.

8.3.5 Cautions

The multiplier's fundamental-frequency rejection depends on balance, especially the phase balance of the input hybrid or balun. The hybrid and balun shown in Figures 8.8 and 8.9 have good phase balance over only a 10–15% bandwidth. Phase error and therefore fundamental-frequency leakage will be worse at the band edges. In most cases that will be the primary bandwidth limitation.

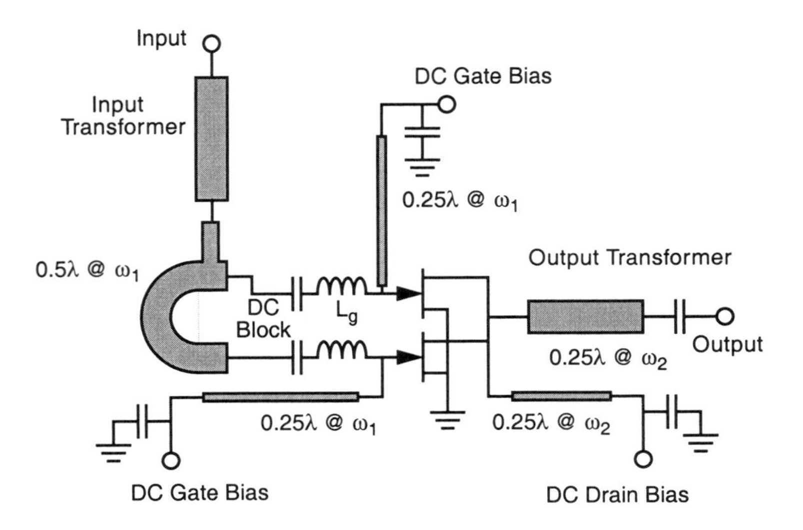

Figure 8.9 A 180-degree hybrid is not always necessary in a balanced multiplier. A balun consisting of a 180-degree loop of transmission line also can be used. It is good for approximately a 15% bandwidth.

The effect of imbalance on fundamental-frequency rejection can be found easily from the following relation:

$$I = -10\log\left(\frac{1 - 2\sqrt{G}\cos\phi + G}{1 + 2\sqrt{G}\cos\phi + G}\right) \tag{8.15}$$

where I is the additional isolation provided by the balanced structure, G is the gain balance, defined as the ratio of the gains through the individual multipliers, and ϕ is the phase balance, defined as the difference of the phase shifts through the individual multipliers. The equation shows that 20 dB of rejection requires maintaining phase balance better than 10 degrees and amplitude balance better than 1 dB. To some degree, amplitude balance can be traded off against phase balance.

The connection between the FETs' drains must be very short, ideally less than 20 degrees of phase at the fundamental frequency, and the point where the output is connected should be precisely halfway between the drains. As with other balanced FET circuits, a long interconnection results in a reactive fundamental-frequency drain termination. This results in feedback and possibly instability.

Because the halfway point between the drains is a virtual ground, the connection from one drain to the halfway point is a shorted stub. The reactance at the drain, X_d, is, therefore,

$$X_d = Z_0 \tan(\theta) \tag{8.16}$$

where Z_0 is the characteristic impedance of the connection line and θ is the length, in degrees of phase, from the drain to the halfway point. Most important, do not allow this reactance to become so great that it resonates the combined drain-to-source capacitances of the FETs at ω_1, or the multiplier may become very unstable.

REFERENCES

[1] Rauscher, C., "High-Frequency Doubler Operation of GaAs Field-Effect Transistors," *IEEE Trans. Microwave Theory Tech.*, vol. MTT-31, 1983, p. 462.

[2] Matthaei, G., L. Young, and E. Jones, *Microwave Filters, Impedance-Matching Networks, and Coupling Structures*, Norwood MA: Artech House, 1980.

[3] Chen, P., C.-T. Li, and P. Wang, "Performance of a Dual-Gate GaAs MESFET as a Frequency Multiplier at Ku-Band," *IEEE Trans. Microwave Theory Tech.*, vol. MTT-27, 1979, p. 411.

[4] Camargo, E., and F. Correra, "A High Gain GaAs MESFET Frequency Quadrupler," 1987 IEEE *MTT-S International Microwave Symposium Digest of Papers*, p. 177.

[5] Stancliff, R., "Balanced Dual-Gate FET Frequency Doublers," 1981 IEEE *MTT-S Int. Microwave Symp. Digest of Papers*, p. 143.

About the Author

Steve Maas received BSEE and MSEE degrees in Electrical Engineering from the University of Pennsylvania in 1971 and 1972, respectively, and a Ph.D. in Electrical Engineering from UCLA in 1984. Since then, he has been involved in research, design, and development of low-noise and nonlinear microwave circuits and systems at the National Radio Astronomy Observatory (where he designed the receivers for the Very Large Array), Hughes Aircraft Co., TRW, the Aerospace Corp., and the UCLA Department of Electrical Engineering. He is now President and Principal Consultant of Nonlinear Technologies, Inc., specializing in the design of low-noise and nonlinear microwave components and systems, and minimization of distortion. He is also Chief Technology Officer of Applied Wave Research, Inc., and is the author of the nonlinear circuit-analysis engine in VoltaireXL®.

Steve is the author of *Microwave Mixers* (Artech House, 1986 and 1992) and *Nonlinear Microwave Circuits* (Artech House, 1988). From 1990 until 1992 he was the editor of the *IEEE Transactions on Microwave Theory and Techniques* and from 1990 to 1993 was an Adcom member and Publications Chairman of the IEEE MTT Society. He received the *Microwave Prize* in 1989 for his work on distortion in diode mixers. He is a Fellow of the IEEE.

Index

A

Active frequency multiplier 241
  balanced 254
  channel-current components 245
  efficiency 242
  feedback 247
  FET 249
  fundamental-frequency rejection 257
  optimization 249
  stability 249, 253
  waveforms 244
Active mixer 175
  conversion efficiency 180
  drain-pumped 181
  dual-gate 190
  FET balanced 199
  single-device equivalent circuit 181
  single-FET 182
  source-pumped 182
AlGaAs 73
Alumina 18, 19

B

Balance 258
Balanced amplifier 44
Balanced mixer
  90-degree hybrid 106
  180-degree hybrid 103
  balun 102
  FET 199, 202
  hybrid 101
  IMD 101
  MOSFET 205
  rat-race 104
Balun 44, 45, 102
Beam-lead diode 57
Beta 63
Bipolar junction transistor (BJT)
  available gain-bandwidth product 67
  base resistance 71
  base-width modulation 71
  beta 63
  capacitance 66
  current gain-bandwidth 66
  Early effect 71
  Gilbert-cell mixer 209
  Gummel-Poon model 68
  high-level injection 69
  I/V characteristic 64
  large-signal equivalent circuit 67
  small-signal equivalent circuit 67
Boltzmann's constant 99
Buffer layer 54
Burckhardt theory 132

C

Capacitor 29
  chip 29
  MIM 30
Characteristic impedance 4, 21
Charge-control model 68
Chip capacitor 29
Circulator 48
Coaxial line 5
  characteristic impedance 7
  cutoff frequency 7
  dispersion 7
  phase velocity 8
Composite material 20
Conversion efficiency 180
  Dual-gate FET mixer 192
  FET mixer 180
  FET multiplier 247, 249
conversion loss 99

resistive multiplier 129
Coplanar multiplier 144
Coplanar waveguide (CPW) 2, 24
Coupled lines 15
Cutoff frequency 58

D

Detector 159
envelope 163, 166
square law 162
Detector diode 161
Dielectric loss 14
Differential mixer 202
Diode mixer 95
180-degree balanced 104
balanced mixer 101
conductance waveform 97
embedding impedance 98
image 100
intermediate frequency 97
intermodulation distortion 100
LO 97
matching 99
mixing 97
mixing product 97
monolithic 123
RF 97
ring mixer 109
spectrum 98
spurious response 100
star 119
Diode selection
rat-race mixer 104
ring mixer 110
Directional coupler 46
Discontinuity 36
Dispersion 22
Doubly balanced mixer
diode mixer 110
horseshoe balun mixer 115
MOSFET 205
ring mixer 109
star 119
star mixer 119
Doubly balanced multiplier 148
Drain-pumped mixer 181
IMD 181
DSB modulator 168
Dual-gate FET 190
Dual-gate FET mixer 190
conversion efficiency 192
Duroid 18

E

Early effect 71
effective dielectric constant 21
envelope detector 159, 163, 166
epitaxial layer 54
even mode 16

F

Feedback in multipliers 247
FET 74
capacitance 78, 83
frequency multiplier 249
gate resistance 82
linear region 76
pinch-off region 76
recessed gate 81
Shockley model 77
source resistance 82
transit time 79
FET acronyms 75
FET frequency multiplier;
see also active frequency multiplier
balanced 254
conversion efficiency 247
device selection 251
fundamental-frequency rejection 257
isolation 249
optimization 249
stability 249
FET mixer 202;
see also active mixer
device selection 184
doubly balanced mosfet 205
dual-gate 190
FET resistive mixer 216
gate-bias circuit 186
input impedance 184
MOSFET 205
output impedance 185
resistive 213
saturation in balanced mixers 207
single-device 182
singly balanced 199
FET resistive mixer 213, 216
bias
conductance waveform 217, 238
dc bias 234
device selection 223
doubly balanced 232
FET model 218
hamster-pumped 215
IF balun 229
impedances 217
isolation 228
JFET and MOSFET 219
LO Power 225
single device 216

single-device equivalent circuit of ring mixer 234
single-device mixer 221
singly balanced 227
subharmonically pumped 236
tuning 225
Field-effect transistor 74
Four-wire IF balun 113
Frequency multiplier 125
resistive 127, 138
resistive (theory) 127
SRD 134
varactor 129
Fundamental-frequency rejection 257
Fused silica 17, 18

G
GaAs MESFET; *see* MESFET
Gate length 76
Gate resistive mixer 182
Gate width 76
Gilbert cell 206, 209
BJT mixer 209
FET mixer 206
used as a frequency multiplier 257
used as a modulator 170
Gummel-Poon model 68

H
Hamster-pumped mixer 215
Heterojunction 87
Heterojunction bipolar transistor (HBT) 52, 73
equivalent circuit 74
structure 74
High-electron-mobility transistor (HEMT) 52, 75, 87
equivalent circuit 89
pseudomorphic 87
structure 88
two-dimensional electron gas 87
Horseshoe balun 115
Hybrid 47, 101
90-degree 47
branch-line 108
rat-race 104
Hybrid coupler 44
Hyperabrupt varactor 62

I
Idler 130
Image 100
IMD 74, 100
Indium phosphide (InP) 18
Inductor 31
Intermediate frequency 97
Intermodulation distortion 74, 100
I-Q modulator 173
Isolation 258
FET multiplier 249, 257
Isolator 48

J
Junction field-effect transistor (JFET) 52, 75
I/V characteristic 77
large-signal equivalent circuit 78
small-signal equivalent circuit 78

L
Lange coupler 47
Linear mixing 213
Linear region 76
LO power 225
diode mixer 99
FET resistive mixer 222, 225
Loss 22

M
Majority-carrier device 53
Marchand balun 119
design 121
MBE 73
MESFET 52, 75, 81
electrical characteristics 84
I_{dss} 85
large-signal equivalent circuit 84
noise figure 85
operation 81
performance 85
R_{ds} 86
small-signal equivalent circuit 85
structure 81
Microstrip 2, 23
Mid-frequency operation 211
MIM capacitor 30
Mixer; *see* individual mixer types
Mode 7
Modulator
DSB 168
Gilbert cell 170
I-Q 173
SSB 171
Molecular-beam epitaxy 73
Monolithic substrate 20
MOSFET 52, 75, 90
electrical characteristic 91
enhancement/depletion mode 75, 90
equivalent circuit 92
I / V characteristic 91
in FET resistive mixer 223
laterally diffused 91, 92
threshold voltage 90
MOSFET mixer 205

N
Normalization resistance 39

O
Odd mode 16
Ohmic contact 54
Oxide layer 54

P
Phase velocity 6, 21
PHEMT; *see* HEMT, pseudomorphic
Pinch-off voltage 75, 76
Planar transmission line 17

Q
Quadrature-coupled amplifier 44
Quasi-TEM line 24

R
Radial stub 35
Rat-race hybrid 143
  broadband 106
  FET frequency multiplier 257
Rat-race hybrid mixer 103
Rat-race multiplier 142
Recessed gate 81
Reflection coefficient 11
Resistive multiplier 127, 139
  balanced 148
  coplanar 144
  diode selection 139
  doubly balanced 148
  singly balanced 141
Resistor 28
Return loss 13
Ring mixer 109
  balun design 111
  diode selection 110
  FET 232
  four-wire IF balun 113
  IF design 113
  RF/LO design 112

S
Sapphire 18, 19
Scattering parameter 37
  conversion to other parameters 41
  two-port 43
Schottky-barrier diode 51
  cross section 54
  cutoff frequency 58
  detector 161
  electrical characteristics 54
  frequency multiplier 127
  junction conductance 99
  measurements 59
  monolithic circuit 58
  practical structures 56
  selection 57
  varactor 61
Schottky-barrier varactor 131
Shockley model 77
Silicon (Si) 18
Single-FET mixer 182
Singly balanced mixer
  90-degree FET 232
  branch-line hybrid mixer 107
  FET 199
  FET differential 202
  FET resistive mixer 227, 231
  rat-race 103
Singly balanced multiplier 142, 144, 145
Solid-state device 51
Source-pumped mixer 182
SPICE 69
Spiral inductor 33
Spurious response 100
Square-law detector 159, 162, 164
  diode selection 165
  sensitivity 163, 165
SRD 52, 61, 134, 154
SRD multiplier 134, 155
  efficiency 136
  pulse generator 135
SSB modulator 171
Standing wave 13, 14
Star mixer 119
  balun design 121
  diode selection 121
  Marchand balun 119
Step junction 36
Step-recovery diode 52, 61, 134;
  see also SRD
Step-recovery diode multiplier 154
Stripline 2, 25
Subharmonically pumped mixer 236
Substrate 17
Suspended-substrate stripline 2, 27

T
Tangential signal sensitivity 163
Telegrapher's equations 9
TEM 7
Termination
  diode mixer 98
  FET mixer 178, 180
  FET multiplier 247
  FET resistive mixer 216
  resistive multiplier 127
  varactor multiplier 130
Thick-film circuit 28
Threshold voltage 75

Transconductance mixer 177
Transmission line 1
  discontinuity 11
  input impedance 12
  loss 14
  propagation constant 14
  standing waves 13
  stub 34
  substrate 17
Triplate 25

V
Varactor 52, 61
  C/V characteristic 130
  dual-mode 62
  equivalent circuit 62
  hyperabrupt 62
  pn-jjunction 131
  punch-through 62
  Schottky barrier 131
Varactor frequency multiplier 129, 151, 152
  Burckhardt theory 132
  diode 131
  idler 130
Vector modulator 173
VSWR 13

W
Wave variable 37
Wilkinson current source 210

The Artech House Microwave Library

Advanced Automated Smith Chart Software and User's Manual, Version 3.0, Leonard M. Schwab

Analysis, Design, and Applications of Fin Lines, Bharathi Bhat and Shiban K. Koul

Analysis Methods for Electromagnetic Wave Problems, Eikichi Yamashita, editor

C/NL2 for Windows: Linear and Nonlinear Microwave Circuit Analysis and Optimization, Software and User's Manual, Stephen A. Maas and Arthur Nichols

Computer-Aided Analysis, Modeling, and Design of Microwave Networks: The Wave Approach, Janusz A. Dobrowolski

Computer-Aided Analysis of Nonlinear Microwave Circuits, Paulo J. C. Rodrigues

Designing Microwave Circuits by Exact Synthesis, Brian J. Minnis

Dielectric Materials and Applications, Arthur von Hippel, editor

Dielectrics and Waves, Arthur von Hippel

Electrical and Thermal Characterization of MESFETs, HEMTs, and HBTs, Robert Anholt

Frequency Synthesizer Design Toolkit Software and User's Manual, Version 1.0, James A. Crawford

Fundamentals of Distributed Amplification, Thomas T. Y. Wong

Generalized Filter Design by Computer Optimization, Djuradj Budimir

GSPICE for Windows, Sigcad Ltd.

HELENA: HEMT Electrical Properties and Noise Analysis Software and User's Manual, Henri Happy and Alain Cappy

HEMTs and HBTs: Devices, Fabrication, and Circuits, Fazal Ali, Aditya Gupta, and Inder Bahl, editors

High-Power Microwaves, James Benford and John Swegle

LINPAR for Windows: Matrix Parameters for Multiconductor Transmission Lines, Software and User's Manual, Antonije Djordjevic, Miodrag B. Bazdar, Tapan K. Sarkar, Roger F. Harrington

Low-Angle Microwave Propagation: Physics and Modeling, Adolf Giger

MATCHNET: Microwave Matching Networks Synthesis, Stephen V. Sussman-Fort

Matrix Parameters for Multiconductor Transmission Lines: Software and User's Manual, A. R. Djordjevic, et al.

Microelectronic Reliability, Volume I: Reliability, Test, and Diagnostics, Edward B. Hakim, editor

Microstrip Lines and Slotlines, Second Edition, K. C. Gupta, Ramesh Garg, Inder Bahl, and Prakash Bhartia

Microwave and Millimeter-Wave Diode Frequency Multipliers, Marek T. Faber, Jerzy Chamiec, Miroslaw E. Adamski

Microwaves and Wireless Simplified, Thomas S. Laverghetta

Microwave Engineers' Handbook, Two Volumes, Theodore Saad, editor

Microwave Materials and Fabrication Techniques, Second Edition, Thomas S. Laverghetta

Microwave and Millimeter Wave Heterostructure Transistors and Applicatons, F. Ali, editor

Microwave and Millimeter Wave Phase Shifters, Volume I: Dielectric and Ferrite Phase Shifters, S. Koul, and B. Bhat

Microwave and Millimeter Wave Phase Shifters, Volume II: Semiconductor and Delay Line Phase Shifters, S. Koul and B. Bhat

Microwave Mixers, Second Edition, Stephen Maas

Microwave Transmission Design Data, Theodore Moreno

Microwave Transition Design, Jamal S. Izadian and Shahin M. Izadian

Microwave Transmission Line Couplers, J. A. G. Malherbe

Microwave Tubes, A. S. Gilmour, Jr.

Microwaves: Industrial, Scientific, and Medical Applications, J. Thuery

Microwaves Made Simple: Principles and Applicatons, Stephen W. Cheung, Frederick H. Levien, et al.

MMIC Design: GaAs FETs and HEMTs, Peter H. Ladbrooke

Modern GaAs Processing Techniques, Ralph Williams

Modern Microwave Measurements and Techniques, Thomas S. Laverghetta

Monolithic Microwave Integrated Circuits: Technology and Design, Ravender Goyal, et al.

MULTLIN for Windows: Circuit-Analysis Models for Multiconductor Transmssion Lines, Software and User's Manual, Antonije R. Djordjevic, Darko D. Cvetkovic, Goran M. Cujic, Tapan K. Sarkar, Miodrag B. Bazdar

Nonuniform Line Microstrip Directional Couplers, Sener Uysal

PC Filter: Electronic Filter Design Software and User's Guide, Michael G. Ellis, Sr.

PLL: Linear Phase-Locked Loop Control Systems Analysis Software and User's Manual, Eric L. Unruh

The RF and Microwave Circuit Design Cookbook, Stephen A. Maas

RF Design Guide: Systems, Circuits, and Equations, Peter Vizmuller

Scattering Parameters of Microwave Networks with Multiconductor Transmission Lines: Software & User's Manual, A. R. Djordjevic, et al.

Solid-State Microwave Power Oscillator Design, Eric Holzman and Ralston Robertson

Terrestrial Digital Microwave Communications, Ferdo Ivanek, et al.

Transmission Line Design Handbook, Brian C. Waddell

TRAVIS Pro: Transmission Line Visualization Software and User's Manual, Professional Version, Robert G. Kaires and Barton T. Hickman

TRAVIS Student: Transmission Line Visualization Software and User's Manual, Student Version, Robert G. Kaires and Barton T. Hickman

Yield and Reliability in Microwave Circuit and System Design, Michael Meehan and John Purviance

For further information on these and other Artech House titles, including previously considered out-of-print books now available through our In-Print-Forever™ (IPF™) program, contact:

Artech House
685 Canton Street
Norwood, MA 02062
781-769-9750
Fax: 781-769-6334
Telex: 951-659
e-mail: artech@artech-house.com

Artech House
Portland House, Stag Place
London SW1E 5XA England
+44 (0) 171-973-8077
Fax: +44 (0) 171-630-0166
Telex: 951-659
e-mail: artech-uk@artech-house.com

Find us on the World Wide Web at: www.artech-house.com